EXPERIMENTS IN THE PURIFICATION AND CHARACTERIZATION OF ENZYMES

A LABORATORY MANUAL

Experiments in the Purification and Characterization of Enzymes

A Laboratory Manual

THOMAS E. CROWLEY
Department of Mathematics and Natural Sciences,
National University, La Jolla, California

JACK KYTE
Department of Chemistry, University of California,
San Diego, La Jolla, California

AMSTERDAM • BOSTON • HEIDELBERG • LONDON
NEW YORK • OXFORD • PARIS • SAN DIEGO
SINGAPORE • SAN FRANCISCO • SYDNEY • TOKYO
Academic Press is an Imprint of Elsevier

ELSEVIER

Academic Press is an imprint of Elsevier
The Boulevard, Langford Lane, Kidlington, Oxford, OX5 1GB, UK
225 Wyman Street, Waltham, MA 02451, USA

First published 2014

British Library Cataloguing in Publication Data
A catalogue record for this book is available from the British Library

Library of Congress Cataloguing in Publication Data
A catalogue record for this book is available from the Library of Congress

ISBN: 978-0-12-409544-1

For information on all Academic Press publications
visit our website at store.elsevier.com

Printed and bound by CPI Group (UK) Ltd, Croydon, CR0 4YY

Working together
to grow libraries in
developing countries

www.elsevier.com • www.bookaid.org

Dedication

Dedicated to Klaus Weber

Contents

Section 1
FNR

1. Purification and Characterization of Ferredoxin-NADP$^+$ Reductase from Chloroplasts
of *S. oleracea*

Section 2
LuxG

Section 3
LDH

Section 4
EXPERIMENTAL DESIGN

4. Experimental Design

List of Figures

SECTION 2: LUXG

SECTION 3: LDH

SECTION 4: EXPERIMENTAL DESIGN

Portions of Figures 1.4, 1.28, 2.5, 3.1, and 3.16 were created using ChemBioDraw® Ultra software, available from CambridgeSoft Corporation, a wholly owned subsidiary of PerkinElmer.

Preface

The exercises in this book were used to teach an upper-division biochemical laboratory course for undergraduate majors in the Department of Chemistry and Biochemistry at the University of California San Diego. Initially, the course focused on the purification and characterization of mammalian lactate dehydrogenase. Later on, two new sets of exercises were created that established a new theme: the comparison of enzymes, across a wide phylogenetic range, catalyzing the oxidation and reduction of cosubstrates with the coenzymes NADH and NADPH. Exercises examining a bacterial FMN reductase and the ferredoxin-NADP$^+$ reductase from a plant expanded the scope of the course. In addition, a section on experimental design has been included. Hypotheses regarding aspects of the enzymes mentioned above or related enzymes are given. Hints are provided to help students design appropriate procedures, perform experiments, and interpret data.

Acknowledgements

We would like to thank Florence F. Davidson, Mark S. Johnson, and Stuart F. Duffy for their contributions to an earlier version of experimental Section 3 while they were graduate students in the department named above. The technical assistance of Peter Wotruba in the same department has been indispensable for teaching the course. He provided valuable advice to new instructors and teaching assistants and prepared all the necessary reagents for the experiments. Other graduate student teaching assistants suggested some refinements and modifications of some of the techniques. Portions of an early version of the manuscript were reviewed by the following scientists and their comments and suggestions were quite valuable: Karl Haushalter of the Department of Chemistry at Harvey Mudd College; Bridget Salzameda of the Department of Chemistry at the University of San Diego; Jodi Kreiling of the Department of Chemistry at the University of Nebraska, Omaha; P. Clint Spiegel of the Department of Chemistry at Western Washington University; Robert M. Bellin of the Department of Biology at the College of the Holy Cross; and Roger K. Sandwick of the Department of Chemistry and Biochemistry at Middlebury College.

Notes Regarding Nomenclature

The names used for the enzymes and metabolites in this book are the "systematic" or "accepted" names recommended by the International Union of Biochemistry and Molecular Biology (IUBMB; http://www.chem.qmul.ac.uk/iubmb/).

Explanations of the Cahn–Ingold–Prelog system (using *R* and *S* prefixes) and the Fischer–Rosanoff system (using D and L prefixes) for naming of enantiomers of chiral molecules are available from the International Union of Pure and Applied Chemistry (IUPAC; http://www.chem.qmul.ac.uk/iupac/).

Abbreviations

LENGTH

cm	centimeter (1×10^{-2} meter)
mm	millimeter (1×10^{-3} meter)
nm	nanometer (1×10^{-9} meter)

MASS

g	gram
mg	milligram (1×10^{-3} g)
μg	microgram (1×10^{-6} g)
ng	nanogram (1×10^{-9} g)

VOLUME

L	liter
mL	milliliter (1×10^{-3} L)
μL	microliter (1×10^{-6} L)

CONCENTRATION

M	molar
mM	millimolar (1×10^{-3} M)
μM	micromolar (1×10^{-6} M)
nM	nanomolar (1×10^{-9} M)

MOLES

mol	mole
mmol	millimole (1×10^{-3} mol)
μmol	micromole (1×10^{-6} mol)
nmol	nanomole (1×10^{-9} mol)

SPECTROPHOTOMETRY AND TIME

A	absorbance
ε	extinction coefficient
l	path length
h	hour
min	minute
s	second

ELECTROPHORESIS

V   volt
A   ampere
mA   milliampere (1×10^{-3} A)

MISCELLANEOUS

g the acceleration of gravity, 9.8 m s^{-2}
rpm   revolutions per minute

Safety Guidelines for Biochemical Laboratories

No eating or drinking is allowed in any chemical laboratory.

ESSENTIAL PERSONAL PROTECTIVE EQUIPMENT FOR ALL PROCEDURES IN A LABORATORY

Eye Protection

Everyone in the laboratory should be wearing eye protection whenever any work with chemicals is being done. For most of the procedures described in this book, *plastic safety glasses* are sufficient. Some procedures involve possible splashing of large volumes of hazardous liquids such as a mixture of isopropanol and dry ice, high-speed rotary devices, or high-frequency vibrational devices. For such procedures, *safety goggles* should be worn.

Laboratory Coat

Everyone in the laboratory should be wearing a laboratory coat whenever any work with chemicals is being done. The sleeves should extend to the wrists.

Gloves

Most of the procedures in this book may be performed safely without gloves. Wear gloves, however, when handling any hazardous chemicals (refer to Table s.1). The chemicals listed in this table are identified as hazardous within the appropriate exercises. If you are allergic to latex, wear nitrile or other hypoallergenic gloves.

TABLE s.1   The Most Hazardous Chemicals to be Used in the Exercises in this Book

Chemical	Procedure(s)	Hazard[a]
Dithiothreitol	Assay for LuxG-his$_6$ activity	Neurotoxin, respiratory tract irritant
NiSO$_4$	Purification of LuxG-his$_6$	Carcinogen, teratogen
Phenylmethanesulfonyl fluoride	Lysis of spinach and bacteria	Toxic to nerves, heart, blood, eyes
Acrylamide	Gel electrophoresis	Neurotoxin, carcinogen, teratogen
N,N,N´,N´-tetramethyl-ethylenediamine	Polymerization of acrylamide	Burns skin and eyes, flammable
Ammonium persulfate	Polymerization of acrylamide	Burns skin and eyes
2-sulfanylethanol (2-mercaptoethanol)	Denaturation of protein	Skin and eye irritant, respiratory tract irritant, mutagen
5,5-diethylbarbituric acid (barbital)	Electrophoresis of the isoenzymes of LDH	Neurotoxin, carcinogen
5-methylphenazinium methyl sulfate	Stain for activity of LDH	Irritant to respiratory tract, eyes and skin
Nitro Blue Tetrazolium	Stain for activity of LDH	Irritant to respiratory tract, eyes and skin
Coomassie Brilliant Blue G-250	Bradford assay, detection of protein in gels	Irritant to respiratory tract, eyes and skin
Phosphoric acid	Bradford assay	Burns eyes, skin, and respiratory tract; toxic to liver, blood, and bone marrow

[a] *Health hazard information from the Material Safety Data Sheet on the Sigma-Aldrich website: http://www.sigmaaldrich.com.*

Proper Footwear

Everyone in the laboratory must wear closed-top shoes that do not have high heels.

HEALTH HAZARDS FROM LABORATORY EQUIPMENT

For all of the equipment described below, obtain instructions from the instructor or the teaching assistant before using the device. You should be supervised during your first use of the device.

Motor-driven Homogenizers

Instructions and photos explaining proper use are in the experimental sections. To ensure your safety, be certain the pestle shaft is tightly secured within the chuck of the drive. During homogenization, the glass receptacle should be held near the top because it is more likely to shatter near the bottom than the top.

Centrifuges

A two-pan balance must be used to bring samples that will be on opposite sides of the rotor to an equal mass before loading. Samples should be in a screw-cap tube or bottle that is **designed for the amount of centrifugal force that will be applied** (photo in the first exercise on ferredoxin-NADP$^+$ reductase). The tubes or bottles will be open while balancing (to allow addition of more solution) but the caps should be on the balance trays. If rotor adaptors are to be used, they should also be on the trays. In most cases, the rotor is only secured to the centrifuge spindle if its cap has been screwed on tightly. After starting the centrifugation, stay near the centrifuge until it reaches the desired rotational speed. If severe vibration of the machine occurs, stop the centrifugation immediately!

Devices for Electrophoresis

Instructions and photos on the proper use of these devices are provided in the experimental sections. Relatively high voltages of direct current are used, and the buffers are excellent conductors. The main concern for safety is that the operator might dip a finger into a buffer tank after the voltage has been applied, and if there is a conduction path through the operator, a severe electric shock might be experienced. The devices described in this manual have *safety interlocks* that are designed so that the electrical circuit is not completed until the lid is covering the buffer tanks. Most electrophoretic apparatuses have a similar design. In spite of this safety feature, have the power supply turned off when the lid is not on and the buffer tanks are not covered.

Emitters of High-frequency Sound

The frequency emitted by a sonifier can damage the human ear. Instructions and photos regarding safe use of this device are provided in the first exercise for LuxG. In particular, the operator should wear ear muffs and eye protection.

Gas Burners and Hot Plates

In preparing a boiling water bath for the denaturation of protein, boiling chips should be used and an electrical hot plate is preferred to a gas burner. When you are finished with the experiment, the hot plate should be turned off and unplugged. The valve at the counter top for any gas burner should be closed immediately after it has been used.

HAZARDOUS CHEMICALS

Definitions

Department of Environmental Health and Safety

At most universities, this department establishes safety regulations for the use of all campus laboratories (it may have a different name in some cases). The staff of the department of environmental health and safety also enforce policies and provide services such as the certification of fume hoods, the disposal of hazardous waste, the maintenance of safety showers and stations for washing eyes, and the response to emergencies. Some colleges may not have a department of environmental health and safety, but all schools with chemistry laboratories will have a safety officer to handle these issues.

Material Safety Data Sheet

The Material Safety Data Sheet (MSDS) provides information on volatility, flammability, toxicity, corrosiveness, and other properties of a hazardous chemical.

Carcinogen

A chemical known to induce mutations that may lead to cancer in the individual that has been exposed.

Concerns for Pregnant Women

Teratogen

A chemical known to induce mutations that may lead to developmental abnormalities in a fetus.

Fetotoxin

A chemical that may be toxic to a fetus.

Information Sources

The Web sites for chemical supply firms usually provide links to the Material Safety Data Sheets for their products. It is also helpful, however, to have access to open-source databases of Material Safety Data Sheets.

Sigma-Aldrich Chemicals:
http://www.sigmaaldrich.com (no subscription necessary).

Where to Find a Material Safety Data Sheet on the Internet (by Interactive Learning Paradigms, Inc.):
http://www.ilpi.com/msds/index.html.

Chemwatch ChemFFX (from Global Chemical Data Inc.):
mailto:info@chemwatchna.com.
http://www.chemwatchna.com/OurServices/MSDSManagement.aspx.

ChemQuik (from Actio Corp., Portsmouth, NH):
mailto:contact@actio.net.
http://www.actio.net/default/index.cfm/products/chemquik/.

Minimization of Exposure

Volatile Chemicals

The laboratory should have a chemical fume hood that is certified at appropriate intervals by the staff of the department of environmental health and safety. Stock bottles and waste containers for volatile chemicals should be kept in the hood.

Corrosive, Toxic or Carcinogenic Chemicals

Wear disposable gloves when handling. If any are spilled on skin, rinse immediately with water.

HAZARDOUS WASTE DISPOSAL

Follow the instructions provided by the instructor or the teaching assistant for disposal procedures for sharp waste, hazardous chemicals, and biohazards (for example, material contaminated with recombinant bacteria).

EMERGENCY RESPONSE

The instructions below for providing immediate assistance are for the instructor, the teaching assistant, or the student. If it appears necessary, a call for professional emergency medical help should be made. Before the first day of class, learn the campus policy covering procedures for such calls. On some campuses, the policy may be to call 9-1-1 directly from the lab. At other schools you may be required to call the department of environmental health and safety or campus health services so that they can call 9-1-1 and direct the responders to the desired location.

Contact of Skin or Eyes with Hazardous Chemicals

The laboratory should have emergency showers and eye-wash stations at the sinks. The instructor, teaching assistants, and students should be aware of their location and how to use them. If necessary, call the *National Capital Poison Center*, 800-822-1222, http://www.poison.org.

Injuries

The laboratory should have a first-aid kit. Everyone working in the laboratory should know where it is located. If an instructor, teaching assistant, or student believes he or she has the necessary understanding to treat a minor injury, they should assist in the treatment.

Sudden Illness such as Fainting or Nausea

Move the affected individual out of the laboratory and away from chemical vapors or other hazards. If the recovery is not rapid and the individual's condition seems serious, call for professional medical help as described above.

Chemical Spills

There should be a *spill kit* in the lab. The instructor or the teaching assistants will instruct you in how it is used. It should contain:

1. Sodium carbonate or sodium bicarbonate for acid spills.
2. Citric acid or sodium bisulfate for alkali spills.
3. Inert adsorbents such as vermiculite, clay, or sand for hazardous liquids.

Fires

The instructor, the teaching assistants, and the students should know the location of the fire extinguishers and be prepared to use them. Learn the evacuation plan on the first day of class.

Earthquakes

If you are in a region prone to earthquakes, learn the recommended response on the first day of class.

General Guidelines for Handling Solutions of Protein

Proteins are generally delicate creatures. While there are cases of proteins that may be boiled for minutes with little or no effect on their activity, a notable case being the family of snake neurotoxins, almost all proteins are inactivated and many precipitate at a much lower temperature. Thermal denaturation is only one of several factors that lead to the death of functional proteins. Some useful precautions are listed below. Following these procedures should allow you to harvest and maintain proteins outside of their normal environment successfully.

1. *Thermal denaturation.*
Keep proteins, tissues, and extracts on ice, refrigerated, or frozen. Repeated cycles of freezing and thawing, however, can produce a loss of function in most proteins. In addition to preventing thermal denaturation, a low temperature decreases the activity of endopeptidases that have been released from their lysosomal cages by cellular disruption. An additional benefit of a lower-than-ambient temperature is the inhibition of the growth of microorganisms.

2. *Water.*
The quality of the solvent, water, used in the preparation of all solutions and the final rinsing of cleaned glassware is crucial. Tap water should never be allowed to come into contact with proteins. A multitude of ions, the effects of which on an enzyme cannot be gauged, and an appreciable quantity of organic solvents and heavy metals are found in tap water. Deionized tap water is processed by a device that combines ion exchange and filtration and usually contains neutral organic chemicals; it is best not to use this type of water. Either water distilled in a glass still or water that has passed through a system that provides similar purity should be used for experiments with enzymes and proteins.

3. *Detergents and soaps.*
Contamination by these two agents is to be avoided at all costs since they cause denaturation. In cleaning your glassware, use these procedures:

   a. Clean with a solution of detergent.
   b. Rinse three times with tap water.
   c. Rinse three times with deionized tap water.
   d. Rinse three times with the water distilled in a glass still (or equivalent).

Three rinses at each step is the minimum necessary to achieve equilibration—remember this when it comes time to rinse a cuvette or other container with buffer before use. Scalpels, scissors, and tweezers should be sterilized with 70% isopropanol, which is a procedure for sterilizing and cleaning that avoids rusting these tools.

4. *Heavy metals.*
The cations of heavy metals such as lead, mercury, and several others can rapidly inactivate a protein. Therefore, the breaking of a thermometer becomes a potentially serious source of

contamination and requires consultation with the instructor with regard to appropriate cleanup and disposal.

5. *Organic solvents.*

Avoid organic solvents; they are another agent for the denaturation of proteins.

6. *Oxidation by air.*

Intracellular enzymes, especially those with an essential cysteine, can sometimes be inactivated through oxidation by the oxygen in the air. This oxidation can be prevented by the addition of 2-sulfanylethanol, glutathione, or dithiothreitol. These agents are rarely used, however, with extracellular proteins, in which cysteines are normally in the form of cystines. Cystine is stable to oxidation by oxygen. These agents, however, interfere with some analytical procedures such as the Lowry method for protein determination.

7. *Foaming.*

If you have ever prepared a meringue for a pie or added egg whites to a margarita for froth, you have observed the effects of the denaturation of the protein in the egg white, predominantly ovalbumin. Protein readily denatures irreversibly at a surface between air and water, and after it has denatured, it becomes a surfactant similar to soap or a detergent. Although a little bit of foaming is not serious and often helps you to locate the protein, a lot of foaming will denature the protein significantly and lower the yield. Avoid foaming. It is often caused by overzealous agitation or stirring of solutions. In this regard, you should be aware that the speed of electric stir motors increases dramatically in the cold room as they warm up. With concentrated solutions of protein, foaming and bubbles are a particular problem when one is using a Pasteur pipette. To avoid this problem, try not to draw more air into the pipette than is necessary.

8. *Fungi and bacteria.*

The growth of bacteria and fungi can be kept at a minimum by keeping solutions of proteins cold, sterilizing scalpels and scissors with 70% isopropanol, and using gloves when handling dialysis tubing. A few drops of isopropanol or toluene are often added as a preservative to the slurries of AMP–Sepharose and Sephacryl. Washing columns of these solid phases before you use them removes these organic solvents.

9. *pH.*

The enzymatic activity of proteins is usually affected by changes in pH, and proteins are unstable and denature at extremes of pH. The behavior of proteins on chromatography and electrophoresis is affected by variations in pH as well. It is for all of these reasons that a buffered solution is used as the solvent for proteins. Buffered solutions, however, have only a finite capacity for absorbing or releasing protons to keep the pH constant. There is always the possibility that one will inadvertently add acid or base in excess of the capacity of the buffered solution. It is always a good idea to check the pH of the solution of protein once in a while, certainly before a major step such as measuring enzymatic activity or loading it onto a chromatographic column or electrophoretic separation. The easiest way to do this is to place a microliter of the solution on a strip of the appropriate pH paper. If you go to your teaching assistant wondering why your protein is inactive, the first question she should ask is, "Have you checked the pH?"

Maintaining a Laboratory Notebook

The ideal notebook for any laboratory course has bound pre-numbered pages. Before each session, you should enter in the notebook a brief summary of the procedures to be performed that day. For some experiments it may be useful to have tables or other formats already laid out for the efficient recording of data. The portions of the notebook that you prepare ahead of time should be neat and well organized.

All of the observations you make during each laboratory session are to be entered into the notebook as you make them. Because you will have to record data and observations quickly, this part of your notebook is not expected to be neat, well organized, or carefully written. In fact, if it is, your teaching assistant will suspect that you were not recording the observations as you made them. You are supposed to be making entries as you work, not after the laboratory is over. You are not to recopy anything in the notebook to make it look better. If any corrections need to be made on a page, this should be done by crossing out erroneous notations or data. Never tear out a page. You should be unconcerned if you spill anything on a page, just wipe it off. The notebooks of Madame Curie have splotches of radium on their pages that are still radioactive today.

You are not allowed to write anything on any piece of paper other than your notebook while you are in the laboratory. Everything that you record while you are in the laboratory is recorded only in your notebook. Every calculation that you make that requires you to write something on a piece of paper will be written in your notebook. You should use a pocket calculator for every calculation.

PERSONAL ELECTRONIC DEVICES

You may not open your laptop computer, talk on a cell phone, listen to music, or turn on any personal electronic device other than your pocket calculator while you are in the laboratory. If possible, you shouldn't even bring any personal electronic device to the laboratory to avoid spilling chemicals on it, especially solutions of electrolytes.

Introduction to Enzymes Catalyzing Oxidation-Reductions with the Coenzyme NAD(P)

PHYSIOLOGICAL ROLE OF NAD(P)

Nicotinamide adenine dinucleotide (NAD) and nicotinamide adenine dinucleotide phosphate (NADP) are carriers of hydride ions that have a central role in the metabolism of all cells (Berg et al., 2012; Voet and Voet, 2010). The only difference in the structures of these two molecules is the phosphate on the number 2′ carbon of the ribose that is bonded to adenine (Figure i.1). They are commonly referred to as *pyridine nucleotides* because the nicotinamide contains the ring of a pyridine. The abbreviation NAD(P) refers to either form. These pyridine nucleotides, by accepting or donating the formal equivalent of a hydride ion, accomplish the respective oxidation–reduction of other substrates. They can also contribute to storage of energy when the oxidized form, NAD$^+$ or NADP$^+$, gains two electrons to become NADH or NADPH, which are the respective reduced forms. They release the stored energy when the reverse reaction occurs and they become oxidized, ultimately by oxygen. The accepted convention for referring to these two molecules without specifying oxidation state is to write the name without the $^+$ or the H; i.e. NAD, NADP, or NAD(P) (Lehninger et al., 2008).

FIGURE i.1 **Oxidation–reduction of nicotinamide adenine dinucleotide and nicotinamide adenine dinucleotide phosphate.** Nicotinamide mononucleotide (top) is linked to adenine mononucleotide (bottom) by an anhydride between the two phospho groups. In nicotinamide adenine dinucleotide (NAD), R at the 2′ position in the ribose of the adenosine is a hydrogen; in nicotinamide adenine dinucleotide phosphate (NADP), R at the 2′ position in the ribose of the adenosine is a phospho group. Because of the phospho groups, the net charge on these molecules is negative. The oxidized form of each, however, has a formal positive charge on the nitrogen of the nicotinamide ring (to the right in the equation) and is therefore denoted NAD$^+$, NADP$^+$, or NAD(P)$^+$. Reduced nicotinamide adenine dinucleotide (phosphate), NADH, NADPH, or NAD(P)H, transfers a formal hydride ion, H$^-$, which is a proton and two electrons (Kyte, 1995, pp 27–30), to another molecule and in the process becomes oxidized nicotinamide adenine dinucleotide (phosphate).

Processes by which a living cell may obtain energy (Lehninger et al., 2008) include:
1. Glycolytic oxidation of glucose (all cell types).
2. Oxidation of pyruvate by the citric acid cycle (aerobic bacteria and mitochondria in eukaryotic cells).
3. Capture of sunlight (photosynthetic bacteria and chloroplasts in algae or plants).

Each of these mechanisms for obtaining energy ultimately results in reduction of $NAD(P)^+$. The reductions of $NAD(P)^+$ accomplished by glycolysis and the citric acid cycle are directly coupled to the oxidations of metabolites that occur within the active sites of enzymes. In photosynthesis, however, a series of electron transport complexes intervene between the capture of the light and the ultimate reduction of $NAD(P)^+$. The energy stored in the reduced NAD(P)H by all organisms is then used in a variety of ways—for example, luminescence in certain bacterial species (Nijvipakul et al., 2008; Winfrey et al., 1997), the synthesis of ATP in mitochondria, and the fixation of carbon in photosynthetic organisms.

COSUBSTRATES FOR NAD(P) IN THE REACTIONS CATALYZED BY FNR, LuxG, AND LDH

The two nicotinamide adenine dinucleotides, NAD and NADP (Figure i.1), are known as *coenzymes*. The term coenzyme has a long and equivocal history. At the moment, a coenzyme is a molecule of low molecular mass that is a substrate for a significant number of enzymes and that transfers a chemical group, hydrogen, or electrons between these different enzymes. A nicotinamide adenine dinucleotide carries the equivalent of a hydride ion between different enzymes. The molecule that, in the active site of an enzyme, receives from or gives to a coenzyme the chemical group, hydrogen, or electrons that it is responsible for transferring is a *cosubstrate*. In the case of NAD or NADP, the metabolite to which the formal hydride ion is added or from which the formal hydride is removed is the cosubstrate. The identity of the cosubstrate is dictated by the particular enzyme that is using the coenzyme. For example, the active site on the enzyme L-lactate dehydrogenase determines that the cosubstrate oxidized by NAD^+ is (S)-lactate and the cosubstrate reduced by NADH is pyruvate.

The exercises described in this manual have as their subjects three enzymes that use NAD(P) as a coenzyme. The processes in which these three enzymes participate are the photosynthetic linear electron transport system, the regeneration of the reduced form of free flavin mononucleotide ($FMNH_2$) for bacterial luminescence, and the cytoplasmic interconversion of lactate and pyruvate. The photosynthetic enzyme (EC 1.18.1.2) is ferredoxin-$NADP^+$ reductase (FNR) from spinach (*Spinacia oleracea*) (Bruns and Karplus, 1995; Negi et al., 2008), the enzyme (EC 1.5.1.42) that regenerates $FMNH_2$ for bacterial luminescence is the FMN reductase (NADH) coded by the *luxG* gene of *Photobacterium leiognathi* (Nijvipakul et al., 2008), and the enzyme (EC 1.1.1.27) that converts lactate and pyruvate is bovine (*Bos taurus*) L-lactate dehydrogenase (LDH) (Eszes et al., 1996; Gulotta et al., 2002). There are many other enzymes that use these pyridine nucleotides as coenzymes. They are members of two classes of enzymes, the dehydrogenases (more than 500 of which use NAD or NADP) and the reductases (more than 100 of which use NAD or NADP), which in turn are both members of the group of enzymes given the name oxidoreductases.

A molecule's ability to be reduced or to reduce another molecule is quantified by its *reduction potential (E)* (Berg et al., 2012; Kyte, 1995, pp 75–80; Lehninger et al., 2008; Voet and Voet, 2010). Reduction potentials are measured in an *electrochemical cell*, with a salt bridge between the reference *half-cell* and the test half-cell. Each half-cell contains both the oxidized and reduced form of the reference or test compound. The potential measured is affected by both the pH in the half-cells and the relative concentrations of oxidized and reduced forms of the molecules being compared. The reference half-cell has a $[H^+]$ of 1 M (pH = 0) and a concentration of H_2 of 1 atmosphere, and the reaction proceeding in the reference cell is the reduction of protons to hydrogen gas. The symbol E would indicate that the test cell is also at pH 0. For biochemistry, it is more practical for the test cell to be at pH 7 and this difference is noted by using the symbol E'. If, in addition to the pH being 7, the concentrations of oxidized and reduced forms of the test compound are equal, the *standard reduction potential* (E'°) is the observed difference in voltage between the two cells.

The reduction potentials are measured in volts and can be positive or negative (Table i.1). The more positive the value, the greater the tendency of the molecule to accept electrons or to be reduced. Although comparison of values of E'° for various molecules is a good way to evaluate their relative affinity for electrons, it is critical to remember that the actual reduction potential (E) of a particular molecule in a living cell will be different if the pH is not 7 and the oxidized and reduced forms are not present in equal concentrations.

TABLE i.1   Standard Reduction Potentials (E'°) for Molecules Involved in Biological Oxidation–Reduction[a]

½ O_2 reduced to one molecule of water	+0.816 volts
FAD reduced to $FADH_2$ in a flavoprotein[b]	−0.040 volts
Pyruvate reduced to lactate	−0.185 volts
FMN reduced to $FMNH_2$, both free in solution[c]	−0.21 volts
NAD^+ reduced to NADH	−0.320 volts
$NADP^+$ reduced to NADPH	−0.324 volts
Ferredoxin (Fe^{3+}) reduced to ferredoxin (Fe^{2+})	−0.432 volts

[a] *Except where noted, all values are from the same reference (Lehninger et al., 2008).*
[b] *The value E'° for FAD varies depending on the protein with which it is associated (Voet and Voet, 2010).*
[c] *(Mayhew, 1999).*

Although not directly relevant to the enzymes addressed here, reduction of molecular oxygen to water is included in the table because it is the ultimate oxidant when the energy deposited in NAD(P)H is withdrawn. The $NAD(P)^+/NAD(P)H$ and O_2/H_2O pairs are at the ends of both the electron transport system in the inner mitochondrial membrane and the photosynthetic linear electron transport chain in the chloroplast. The values of E'° for these pairs are indicative of their role in these processes. In the mitochondrion, electrons flow from NADH through various carriers and eventually reduce O_2 to H_2O. The more positive value of E'° for O_2 explains why the flow in this direction releases energy that leads to ATP synthesis. The direction of flow is energetically favorable. In contrast, electron flow in the photosynthetic linear electron transport chain from H_2O to $NADP^+$ is not energetically favorable (Barber, 2009; Shikanai, 2007). Consequently, the input of energy in quanta of light is necessary to drive the electrons in this unfavorable direction.

FNR: A portion of the light energy captured by photosynthetic linear electron transport is ultimately used to reduce Fe^{3+} in the iron–sulfur protein ferredoxin (Figure i.2A) to Fe^{2+} (Binda et al., 1998). Ferredoxin-NADP$^+$ reductase then catalyzes the transfer of electrons from ferredoxin to NADP$^+$. Unlike the other two enzymes that are subjects of the exercises in this manual, FNR is a *flavoenzyme* (Onda and Hase, 2004). The flavin adenine dinucleotide (FAD) tightly bound to the enzyme is a prosthetic group in its active site (Kyte, 1995, pp 72–90). A *prosthetic group* is a molecule that is stably associated with an enzyme. Electrons are passed from ferredoxin to the prosthetic FAD and then to NADP$^+$.

LuxG: Bacterial luminescence is produced by an enzyme that uses the reducing power of FMNH$^-$, a molecule in free solution in the cytoplasm (Figure i.2B), in a reaction that reduces O_2 to water. The anion FMNH$^-$ is the conjugate base of $FMNH_2$, a molecule that ionizes with a pK_a of 6.3 (Song et al., 2007b). The oxidized coenzyme must be reduced back to FMNH⁻ to support continuous luminescence (Nijvipakul et al., 2008). The oxidation of glucose or other nutrients by glycolysis and the citric acid cycle within the bacteria produces NADH. The reduction of FMN by NADH, catalyzed by the FMN reductase encoded by the *luxG* gene, provides a steady stream of FMNH⁻.

FIGURE i.2   **Oxidation–reduction of the cosubstrates for NAD(P) in the reactions catalyzed by FNR and LuxG.** (A) A representation of a [2Fe–2S] iron–sulfur cluster, like the one in ferredoxin. Each iron atom is in a tetrahedral coordination with two inorganic sulfides and the thiolates from two cysteines for a total formal charge of −8 in the complete set of ligands. The two irons are both 3+ in the oxidized state or both 2.5+ in the reduced state, the latter because they are electronically coupled. The reduced state in the drawing, however, is represented by having one 3+ and the other 2+. Each molecule of ferredoxin contains one of this type of cluster and can carry one electron. Ferredoxin-NADP⁺ reductase catalyzes the transfer of one electron from each of two of these clusters on separate molecules of ferredoxin, one after the other, to one molecule of the prosthetic FAD. The prosthetic FAD, because it can exist in the stable semiquinone, can be reduced by one electron at a time. As a result, it can gather one at a time the two electrons needed to reduce the NADP⁺. (B) The isoalloxazine of flavin formed from three fused six-membered rings is the site of oxidation–reduction in the prosthetic FAD in the active site of FNR and in the substrate FMN for LuxG. In FAD, R is an adenosine monophosphate, whereas in FMN it is a hydrogen atom. Although the reduced forms are usually shown as FADH₂ and FMNH₂, at pH greater than 6.3 these reduced flavins are ionized to FADH⁻ and FMNH⁻ (Song et al., 2007b).

LDH: During glycolysis, two molecules of NADH are produced from each molecule of glucose. Under conditions of moderate to severe anaerobiosis, when the cell is obtaining most if not all of its energy from glycolysis, these two molecules of NADH must be reoxidized to NAD⁺. Otherwise, glycolysis would run out of NAD⁺. Lactate dehydrogenase takes the pyruvate produced by glycolysis and uses it to reoxidize the NADH to allow glycolysis to proceed continuously (Figure i.3). Under these conditions, the product of this anaerobic glycolysis, balanced in its oxidation and reduction, is lactate. The central carbon of lactate is a chiral center. Both (*S*)- and (*R*)-lactate are of biological importance and reactions involving the two isomers are catalyzed by distinct enzymes: L-lactate dehydrogenase and D-lactate dehydrogenase (cytochrome) (EC 1.1.2.4), respectively. The exercises in the experimental section titled "LDH" use bovine L-LDH. For the remainder of this book, unless stated otherwise, "lactate" will refer to (*S*)-lactate and "LDH" will refer to L-lactate dehydrogenase.

FIGURE i.3 **The cosubstrates of NAD in the reactions catalyzed by LDH.** Pyruvate is reduced by NADH to (*S*)-lactate. Using the Cahn–Ingold–Prelog system, the systematic name for this enantiomer of lactate is (*S*)-lactate. This is the name recommended by the International Union of Biochemistry and Molecular Biology, and for this reason, it is used here. The name L-lactate (Fischer–Rosanoff system), however, is also an accepted name for this molecule.

When NAD$^+$ is needed to maintain glycolytic rate, NADH reduces pyruvate to lactate. When pyruvate is needed to feed the citric acid cycle, electron flow from lactate to NAD$^+$ oxidizes lactate to give pyruvate. The standard reduction potentials of these redox pairs (Table i.1) suggest that when the oxidized and reduced forms of each of these molecules are equally abundant, the pyruvate-to-lactate conversion is energetically favored (pyruvate has greater affinity for electrons than does NAD$^+$). When lactate, however, becomes more abundant than pyruvate or NAD$^+$ becomes more abundant than NADH or both of these situations occur simultaneously, the conversion of lactate to pyruvate will be favored (Lehninger et al., 2008; Voet and Voet, 2010). These are the conditions that occur when the cell is receiving O$_2$ and respiration is active. In this situation, both pyruvate (for the citric acid cycle) and NADH (for the mitochondrial electron transport chain) are flowing into mitochondria.

DOES FNR FUNCTION AS A MONOMER OR A DIMER?

That mammalian LDH is a tetramer is well established (Eszes et al., 1996; Read et al., 2001). Convincing evidence that LuxG is a homodimer has also been presented (Nijvipakul et al., 2008). For FNR, however, there is evidence that the enzyme can function as either a monomer or a dimer.

Although it functions in the chloroplast, FNR from *S. oleracea* is encoded on a nuclear gene and is synthesized on cytoplasmic ribosomes (Onda and Hase, 2004). A transit peptide 55 amino acids in length is removed from the amino terminus during translocation into the chloroplast. Because of the nature of this process of translocation, dimerization can occur only in the chloroplast after translocation has been completed (Negi et al., 2008). Evidence has been presented, however, suggesting that some of the folded, enzymatically active polypeptides of FNR remain as monomers in the chloroplast (Zhang et al., 2001). The monomers are believed to be associated with plastoquinol–plastocyanin reductase (the same enzyme is also known, for historical reasons, as the cytochrome b_6f complex) and have a role in cyclic electron transport (Shikanai, 2007) rather than linear electron transport.

The membrane of the thylakoid assumes a convoluted shape within the chloroplast and harbors the protein and lipid components of the linear electron transport chain (Lehninger et al., 2008). The thylakoid is analogous to the inner membrane of a mitochondrion. The lumen is the interior of the compartment bounded by the membrane of the thylakoid while the stroma is the remainder of the chloroplast outside of the membrane of the thylakoid. Although most scientists studying the role of FNR in linear electron transport believe the enzyme is associated with the stromal side of the thylakoid membrane, the sequence of amino acids does not show a transmembrane domain that might anchor it there and a mechanism for binding of the enzyme to the membrane has not been elucidated. Experiments in spinach (Zanetti and Arosio, 1980) and the small flowering plant *Arabidopsis thaliana* (Lintala et al., 2007) have led to the suggestion that dimerization is critical for membrane localization.

Because it is an important aspect of the role of FNR in photosynthesis and it is not yet completely resolved, it is worthwhile to address the issue of whether the enzyme is a monomer or a dimer in different situations. There are two methods typically applied to a purified enzyme for this purpose: electrophoresis on gels of polyacrylamide cast in solutions of sodium dodecyl sulfate in concert with chromatography of the native enzyme by molecular exclusion (also called gel filtration) and quantitative cross-linking. In the former, the electrophoresis elucidates the length of the polypeptide that constitutes the subunit, while the gel filtration of the nondenatured enzyme measures the surface area of the native protein (Kyte, 2007, pp 421–431). In the latter, the molecules of protein are cross-linked by a reagent that covalently connects all of the subunits in an oligomer without cross-linking oligomers to each other, and the size of the cross-linked product is determined by electrophoresis on gels of polyacrylamide cast in solutions of sodium dodecyl sulfate (Kyte, 2007, pp 439–446). Such experiments are described in the FNR section and the results will provide an estimate of the number of subunits in the enzyme.

DO FNR, LuxG, AND LDH HAVE DIFFERENT MICHAELIS CONSTANTS FOR NAD(P)?

One of the parameters describing the kinetics of an enzyme is the *Michaelis constant*, K_m, for one of its substrates (Table i.2) (Berg et al., 2012; Kyte, 1995, pp 149–174; Lehninger et al., 2008; Voet and Voet, 2010). Exercises in this manual guide you through kinetic measurements to determine the values of K_m for the three enzymes to be studied.

TABLE i.2   Symbols and Units used in Enzyme Kinetic Studies

Symbol	Definition	Typical Unit
$[E]_t$	Total concentration of active sites	nM or μM
v	Initial rate or initial velocity	μM s^{-1}
V	Limiting rate or limiting velocity	μM s^{-1}
k_0	Catalytic constant or turnover number	s^{-1}
k_A	Specificity constant for reactant A	μM^{-1} s^{-1}
K_{mA}	Michaelis constant for reactant A	μM or mM

Enzymes, because they are catalysts present in solution in much lower molar concentration than their substrates, cannot alter the equilibrium constant for the chemical reactions that they catalyze, because that would violate the second law of thermodynamics. Consequently, all enzymes necessarily catalyze the reactions for which they are responsible in both directions. When the kinetics of an enzymatic reaction are being studied, the observer chooses to follow the rate of the enzymatic reaction in only one of those two directions. This is accomplished by adding to the solution the complete set of substrates on only one side of the equilibrium describing the enzymatic reaction. The substrates in this set are the *reactants* for the kinetic study. The experiment is performed by mixing all of the reactants with the enzyme. One of the reactants or the enzyme itself is chosen to be the last participant to be added to the mixture. As soon as the last participant is added to initiate the reaction, the concentration of one of the reactants or one of the products is monitored as a function of time. The rate at which the concentration of the monitored reactant decreases, or the rate at which the concentration of the monitored product increases, is the rate of the reaction.

As the concentrations of the products increase over the course of the measurement, the reverse reaction catalyzed by the enzyme increases in rate, and the rate being monitored decreases as the concentrations of reactants and products reach equilibrium. At equilibrium, all measurable changes cease, even though the enzyme is still catalyzing the reaction rapidly in both directions. No further changes occur because at equilibrium the enzyme catalyzes the reaction at the same rate in both directions. To avoid this decrease in the rate of the reaction caused by the increase in the rate of the back reaction, the overall rate of the reaction is only monitored at short times to obtain the initial rate of the reaction, or the *initial velocity* (v).

Initially, as the measurements are being made, the initial velocity is recorded in units of micromoles minute^{-1} milliliter^{-1}. If the concentration of active sites in the solution being monitored is known, dividing the initial velocity in micromoles minute^{-1} milliliter^{-1} by the molar concentration of active sites in micromoles milliliter^{-1} converts the initial velocity to the units of seconds^{-1}. The choice of these units gives the number of times an active site on the enzyme converts reactants to products each second under the conditions of the assay.

One would expect the rate of a reaction catalyzed by an enzyme to increase as concentration of one of the reactants is increased. It is not, however, a simple linear relationship. It is always the case that as the concentration of one of the reactants in an enzymatic reaction increases, the initial velocity increases but it increases less and less until no further increase occurs (Figure i.4) or, in some cases, until the rate begins to decrease. In the former instance, the rate of the reaction reaches saturation; in the latter case, the reactant itself is interfering with the enzymatic reaction as an inhibitor, and its inhibition increases as its concentration increases. The former behavior, saturation, is more common and is indicative of enzymatic reactions. When it occurs, inhibition by a reactant is in addition to the underlying saturation that is occurring.

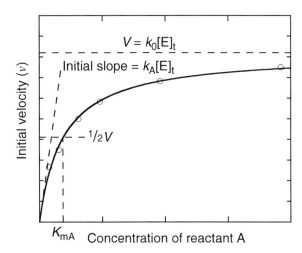

FIGURE i.4 **Kinetic data for an enzyme.** The data chosen for the example are for the hydrolysis of sucrose by β-fructofuranosidase (Michaelis and Menten, 1913; Michaelis et al., 2011). As the concentration of reactant A in the assays is increased, the initial velocity, v, increases and approaches, as a horizontal asymptote, the limiting velocity (V). Initial velocity is typically recorded in units of micromoles minute^{-1} milliliter^{-1} but can be changed to the units of second^{-1} if the molar concentration of active sites is known. The Michaelis constant for reactant A, K_{mA}, has the same units as those used for the concentration of the reactant A.

In the usual, ideal case, the increase in the initial velocity of an enzymatic reaction with increasing concentrations of one of its reactants is described by the equation for a rectangular hyperbola. There are two equations that can be used to describe this particular rectangular hyperbola:

$$v = \frac{(k_0[E]_t)\,(k_A[E]_t[A]_0)}{k_0[E]_t + k_A[E]_t[A]_0} \tag{i.1}$$

$$v = \frac{V[A]_0}{K_{mA} + [A]_0} \tag{i.2}$$

where k_0 is the *catalytic constant*, k_A is the *specificity constant* for reactant A, $[E]_t$ is the total molar concentration of active sites on the enzyme in the solution, $[A]_0$ is the initial molar concentration of reactant A, V is the *limiting velocity*, and K_{mA} is the *Michaelis constant* for reactant A. You should convince yourself that these are two ways of writing the same equation. If the initial velocity of the enzymatic reaction as a function of the concentration of one or more of the reactants (a measurement you will be making in several of the exercises) shows saturation and ideal hyperbolic behavior, the data can be fit by either one of these equations. The meaning of these parameters and the derivation of kinetic equations will be one of the subjects of your lecture courses. In this course, you will be making measurements, showing that the measurements can be described by particular equations, fitting the equations to the measurements, and obtaining numerical values for the parameters of the equations.

At the moment, we are concerned with the observed, empirical parameter, K_{mA} for reactant A, the concentration of which you will be varying in kinetic experiments. Aside from the theoretical meaning of K_{mA}, which will be the subject of other courses, there is an obvious practical meaning of K_{mA}. If you examine the equation defining K_{mA} (Equation i.2), you will see that K_{mA} is equal to the concentration of reactant A at which the initial velocity of the enzymatic reaction is half of its initial velocity at saturation, V. Saturation occurs when the initial concentration of reactant A is infinite. What this means in practice is that in the cell, the concentration of reactant A should be greater than K_{mA} for the enzyme to be operating efficiently; in other words, for most of the molecules of enzyme to be working at the same time. Usually, natural selection has adjusted K_{mA} to be in the range of the normal concentration of reactant A in the cells in which the enzyme is found. Therefore, K_{mA} has practical consequences.

The best way to fit the second equation describing ideal behavior to the data that you will be gathering, in order to obtain a Michaelis constant, is a nonlinear least squares fit. This may be accomplished by many applications. For example: *Microsoft Excel* with the *Solver* add-in (Barton, 2011; Kemmer and Keller, 2010) or *Kaleidagraph* from Synergy Software (www.synergy.com).

In the exercises for this course, we will be studying enzymes that catalyze the following reactions (Equations i.3–i.5):

$$2 \text{ oxidized ferredoxin} + \text{NADPH} \rightleftharpoons 2 \text{ reduced ferredoxin} + \text{NADP}^+ + \text{H}^+ \tag{i.3}$$

$$\text{FMN} + \text{NADH} + \text{H}^+ \rightleftharpoons \text{FMNH}_2 + \text{NAD}^+ \tag{i.4}$$

$$\text{pyruvate} + \text{NADH} + \text{H}^+ \rightleftharpoons (S)\text{-lactate} + \text{NAD}^+ \tag{i.5}$$

In Equation i.5, the *systematic* name for the lactate enantiomer using the Cahn–Ingold–Prelog system and recommended by the International Union of Biochemistry and Molecular Biology,

(S)-lactate, is used. L-Lactate (Fischer–Rosanoff system), however, is also an accepted name for this molecule.

The kinetics of these reactions can be measured in either direction. In the exercises you will be performing, they will be measured in the direction as written, that of the oxidation of NAD(P)H. For the measurements you will be making, the concentration of one of the reactants in each equation will be held constant (usually at a high enough concentration to saturate the enzyme) and the concentration of the other, reactant A, will be varied to obtain numerical values for V and K_{mA}.

Source of Conventions for Discussion of Enzyme Kinetics

International Union of Biochemistry and Molecular Biology
Web site: http://www.chem.qmul.ac.uk/iubmb/kinetics/

Computational Techniques for Biochemistry

SOFTWARE FOR ANALYSIS OF DATA

Software for spreadsheets is an excellent tool for analysis of data. In addition to the widely used *Excel* software from Microsoft, there are several free (or inexpensive) software programs that function in a similar manner. For example:

Open Office Calc (part of the *Open Office* suite from Apache, free download).
Current version (3.4) runs on either Windows or Mac operating systems.
http://www.openoffice.org.

iWork Numbers (from Apple Computers).
Current version (Numbers '09) cost $20. Only runs on Mac operating system.
http://www.apple.com/iwork/numbers.
In the user's guide, the terms *spreadsheet* and *table* appear to be used interchangeably.

Google Docs.
1. Create personal account (free).
2. Create spreadsheets, do calculations, and plot data online.
3. Save files in online database.
http://www.docs.google.com.

For most of the analysis of the data that is necessary for the exercises in this book, the plotting tool referred to as "*x-y* scatter plot" should be used. For the data from gel electrophoresis, a semilog *x-y* scatter plot is needed. All the software mentioned above performs these functions.

COMPUTATIONAL METHODS AND BIOINFORMATICS FOR STUDYING THE STRUCTURE AND FUNCTION OF PROTEINS

Sources of Software for Research in the Life Sciences
A variety of software that runs on desktop PCs is available for computational biochemistry.

Software that May be Installed on Personal Computers.

GeneDoc.
Editing and alignment of the sequences for proteins and nucleic acids.
Visualization of the secondary structures of proteins.
Shading of alignment and secondary structure.

Created by Karl B. Nicholas (Nicholas et al., 1997) and currently maintained by Hugh B. Nicholas, Jr., of the National Resource for Biomedical Supercomputing (NRBSC; associated with the Pittsburgh Supercomputing Center http://www.psc.edu). A free download is available at: http://www.nrbsc.org/gfx/genedoc/index.html.

Vector NTI Advance (from Invitrogen).
   Characterization of the primary, secondary, and tertiary structures of proteins and nucleic acids. A free 30-day trial license is available at:
   http://www.invitrogen.com/site/us/en/home/Products-and-Services/Applications/Cloning/vector-nti-software.html.

BioEdit.
   Editing and alignment of the sequences of proteins and nucleic acids. Created by Tom Hall while at North Carolina State University.
   A free download is available at: http://www.mbio.ncsu.edu/BioEdit/bioedit.html.

Software Used on Websites where Download Might be Allowed.

   National Center for Bioinformatics (NCBI).
      http://www.ncbi.nlm.nih.gov/guide/data-software/ - downloads.

   Swiss Institute of Bioinformatics.
      http://www.isb-sib.ch/services/-software-tools.html.

   Molecular Modeling Database (MMDB) of the NCBI.
      http://www.ncbi.nlm.nih.gov/Structure/MMDB/mmdb.shtml.
      Programs for the visualization and comparison of the tertiary structures of proteins.

Databases for Research in the Life Sciences

Macromolecules.

   Comprehensive:
      National Center for Bioinformatics (NCBI, http://www.ncbi.nlm.nih.gov, includes taxonomy).
      ExPASy of the Swiss Institute for Bioinformatics http://www.expasy.org.

   Primarily for studies of proteins:
      Universal Protein Resource (UniProt, http://www.uniprot.org): a collaboration between the Protein Information Resource (PIR, Georgetown University, Washington D.C.), the European Bioinformatics Institute (EBI) and the Swiss Institute for Bioinformatics. This database includes taxonomy.

   Tertiary structures of proteins:
      Molecular Modeling Database of the NCBI (http://www.ncbi.nlm.nih.gov/sites/entrez?db=structure).
      Protein Data Bank of the Research Collaboratory for Structural Bioinformatics (http://www.rcsb.org/pdb/home/home.do). Managed by UC San Diego and Rutgers University.

Small Organic Molecules.

Free:
PubChem from the NCBI (http://pubchem.ncbi.nlm.nih.gov).
ChemSpider from the Royal Society of Chemistry, Cambridge (http://www.chemspider.com).
National Institute of Standards and Technology Chemistry WebBook (http://webbook.nist.gov/).

Require institutional subscription:
Knovel (http://www.knovel.com/web/portal/main).
The Merck Index (http://www.cambridgesoft.com/databases/themerckindex/default.aspx).
Chem ACX (http://www.cambridgesoft.com/databases/chemacx/default.aspx). Includes suppliers of each listed chemical.

COMPUTATIONAL EXERCISE 1

Retrieve Entries from a Database and Compare Sequences of Amino Acids

This exercise will employ databases and programs available from the NCBI (steps 1–3) and the Universal Protein Resource (UniProt, step 4) to align sequences of amino acids (Kyte, 2007, pp 346–361).

1. Establish a free NCBI account.

Create a user name and password. Although not essential for using the program, it is very helpful for maintaining *Collections* of the sequences of amino acids for proteins and files of structures. This minimizes repetitive searching of the database and allows the user to work more efficiently.

2. Search for the relevant sequences of amino acids (primary structures).

a. Login to your account by clicking *My NCBI* in the upper right. The terms *protein* and *enzyme* may refer to a single polypeptide chain, folded or unfolded, or an oligomeric assembly of folded polypeptides. Each folded polypeptide in a oligomeric complex is usually called a *subunit* but the term subunit is sometimes used with other meanings. Formally, the term subunit means any member of a set of equal portions of the complete molecule. These oligomeric complexes (Kyte, 2007, pp 466–485) may be *homooligomers* (oligomers of identical subunits) or *heterooligomers* (oligomers with two or more different types of subunits). In most protein databases, each entry is for a single polypeptide, but it is crucial to keep in mind that the protein or enzyme that exists in nature may be an oligomer of several folded polypeptides.

Although there is evidence that each of the three enzymes to be studied here exists as an oligomer, only L-lactate dehydrogenase (LDH) is known to have multiple *isoenzymes*. Isoenzymes are two or more molecules of enzyme that are closely related to each other but not identical, that each catalyze the same chemical reaction, and that are all found in the same organism (Kyte, 2007, pp 358–359). The fundamental unit of a homooligomeric enzyme is the subunit. Each subunit is formed from one folded polypeptide. Two subunits are related to each other isoenzymatically if they both catalyze the same reaction, the sequences of their polypeptides are different but clearly related to each other, and they are both found in the same organism. The different polypeptides that fold to create the respective isoenzymatic subunits each have a different sequence of amino acids, but each of these sequences is homologous to the sequences of the other isoenzymatic polypeptides and all are obviously descended from the same common ancestor that existed recently in evolutionary time.

Two isoenzymes can each be homooligomeric if one is composed completely of one type of isoenzymatic subunit and the other completely of another type of isoenzymatic subunit. If two different isoenzymatic subunits can enter the same oligomer interchangeably, however, an isoenzyme can also be created by mixing different isoenzymatic subunits in the same oligomer. Each of the constituent isoenzymatic subunits catalyzes the same reaction and contains an active site almost identical to the active sites in the other types.

The exercises in this manual address both the A (dominant in skeletal muscle) and B (dominant in heart) isoenzymatic subunits of LDH. In the case of LDH, A subunits and B subunits can together enter the same tetramer interchangeably. Although the A and B designations (Goldberg et al., 2010) are currently recommended for these two isoenzymatic subunits, for many years the A isoenzymatic subunit was designated M, for skeletal muscle, and the B isoenzymatic subunit was designated H, for heart. The M and H nomenclature appears in many journal articles (Read et al., 2001) and books. It is important to remember, however, that the A (or M) isoenzyme is expressed in tissues other than skeletal muscle and the B (or H) isoenzyme is expressed in tissues other than heart, which is one of the reasons the nomenclature was changed.

b. The NCBI maintains many databases. Select the *Protein* database and search for the four types of polypeptides that are the subjects in this laboratory manual: those that fold to produce the enzymes FNR from *Spinacia oleracea* (spinach), LuxG from *Photobacterium leiognathi* and the A and the B isoenzymatic subunits of bovine (*Bos taurus*) LDH (Table c.1). In addition to the sequence of amino acids, each entry will indicate who submitted the sequence, and include literature references and taxonomic information. An example, for the *lac* repressor of *Eschericia coli*, is shown (Figure c.1). Create a new *Collection* with the database entries for these four polypeptides.

TABLE c.1   Enzymes Studied in the Exercises in this Manual

Full Accepted Name	Abbreviation	Species	
		Common Name	Phylogenetic Name
Ferredoxin-NAD(P)⁺ reductase	FNR	Spinach	*Spinacia oleracea*
FMN reductase (NADH)	LuxG	Bioluminescent bacteria	*Photobacterium leiognathi*
ʟ-lactate dehydrogenase A isoenzyme	LDH-A	Cattle	*Bos taurus*
ʟ-lactate dehydrogenase B isoenzyme	LDH-B	Cattle	*Bos taurus*

3. Alignments of the sequences of amino acids using the Basic Local Alignment Search Tool (BLAST) of the NCBI.

You will compare the sequences of the amino acids for the enzymes studied in this manual.

a. Open your collection of enzyme database entries and click *Run BLAST*.

b. The BLAST programs are most frequently used for searching an entire database or a subset of a database. In such searches, a *Search Set* must be specified (Table c.2). In this exercise, however, you will be comparing only the four sequences for which you searched earlier. These four are the *Subject Sequences*. You should select the *Align Two or More Sequences* option.

c. Do pairwise comparisons by placing one of the four proteins in the Query field and one of the others in the Subject Sequence field. Click on BLAST to get the result. Repeat with a different protein in the Query field until all combinations have been evaluated.

lac repressor [Escherichia coli UTI89]

NCBI Reference Sequence: YP_539404.1

FASTA Graphics

Go to:⊡

```
LOCUS           YP_539404                363 aa              linear   BCT 27-SEP-2012
DEFINITION      lac repressor [Escherichia coli UTI89].
ACCESSION       YP_539404
VERSION         YP_539404.1  GI:91209418
DBLINK          Project: 58541
                BioProject: PRJNA58541
DBSOURCE        REFSEQ: accession NC_007946.1
KEYWORDS        .
SOURCE          Escherichia coli UTI89
  ORGANISM      Escherichia coli UTI89
                Bacteria; Proteobacteria; Gammaproteobacteria; Enterobacteriales;
                Enterobacteriaceae; Escherichia.
REFERENCE       1  (residues 1 to 363)
  AUTHORS       Chen,S.L., Hung,C.S., Xu,J., Reigstad,C.S., Magrini,V., Sabo,A.,
                Blasiar,D., Bieri,T., Meyer,R.R., Ozersky,P., Armstrong,J.R.,
                Fulton,R.S., Latreille,J.P., Spieth,J., Hooton,T.M., Mardis,E.R.,
                Hultgren,S.J. and Gordon,J.I.
  TITLE         Identification of genes subject to positive selection in
                uropathogenic strains of Escherichia coli: a comparative genomics
                approach
  JOURNAL       Proc. Natl. Acad. Sci. U.S.A. 103 (15), 5977-5982 (2006)
   PUBMED       16585510
ORIGIN
        1 mvnvkpvtly dvaeyagvsy qtvsrvvnqa chvsaktrek veaamaelny ipnrvaqqla
       61 gkqsllligva tsslalhaps qivaaiksra dqlgasvvvs mversgveac kaavhnllaq
      121 rvsgliinyp lddqdaiave aacanvpalf ldvsdqtpin siifshedgt rlgvehlval
      181 ghqqiallag plssvsarlr lagwhkyltr nqiqpiaere gdwsamsgfq qtmqmlnegi
      241 vptamlvand qmalgamrai tesglrvgad isvvgyddte dsscyipplt tikqdfrllg
      301 qtsvdrllql sqgqavkgnq llpvslvkrk ttlppntqta spraladslm qlarqvsrle
      361 sgq
//
```

FIGURE c.1   The entry in the NCBI database for the polypeptide of the *lac* repressor from *E. coli.*

TABLE c.2 Definitions for Use of the NCBI BLAST

Accession number	An identifier for an entry in the database; usually a combination of numbers and letters
blastp	Program for comparing the sequences of amino acids in proteins (blastn is for nucleotide sequences)
Query sequence	The single sequence to be compared to another sequence or sequences; the actual sequence may be entered or may be specified by an *accession number*
Subject sequence(s)	The sequences to be compared to a query sequence; as with the query sequence, the actual sequence(s) may be entered or they may be specified by an accession number(s)
Search set	An entire database of sequences of amino acids, or a subset of such a database (e.g. only those from a particular organism)

4. Alignments of the sequences of amino acids using the alignment program of UniProt.

  a. Go to the address http://www.uniprot.org/.

  b. Find the identification number for each of the four sequences by typing in their names and species in the *Query* box and then choosing the proper sequence from the list that appears and clicking on it. You should look over the information about each one that is provided on its individual page.

  c. Once you have each of the four identification numbers, click on the *Align* option.

  d. First, perform the six pairwise alignments for the four sequences by entering two of the numbers, separated by a return, into the box and clicking *Align*. Repeat these commands five more times, each time for another of the pairwise alignments.

  The final alignments give percent identity and the number of similar positions.

  e. Compare the alignments with this algorithm to the alignments made with BLAST.

5. Interpret results.

  An example of an alignment performed by BLAST for the *lac* repressor protein from two species of bacteria is shown (Figure c.2). The output includes:

a. *Percent identities.*
    These are positions at which the same amino acid occurs.

b. *Percent positives.*
    Usually referred to as *conservative substitutions*. A conservative substitution identifies a position at which the two proteins have different amino acids but the side chains of these amino acids have similar characteristics; for example, they are both nonpolar, both polar, both negatively charged, and so forth. You should be aware that these groupings are subjective.

c. *Alignments of pairs of sequences.*
    Shows identities and conservative substitutions. Gaps are often introduced in one or the other of the sequences to maximize the number of matches over the entire alignment. Each of these gaps is at a position at which natural selection has excised at this location

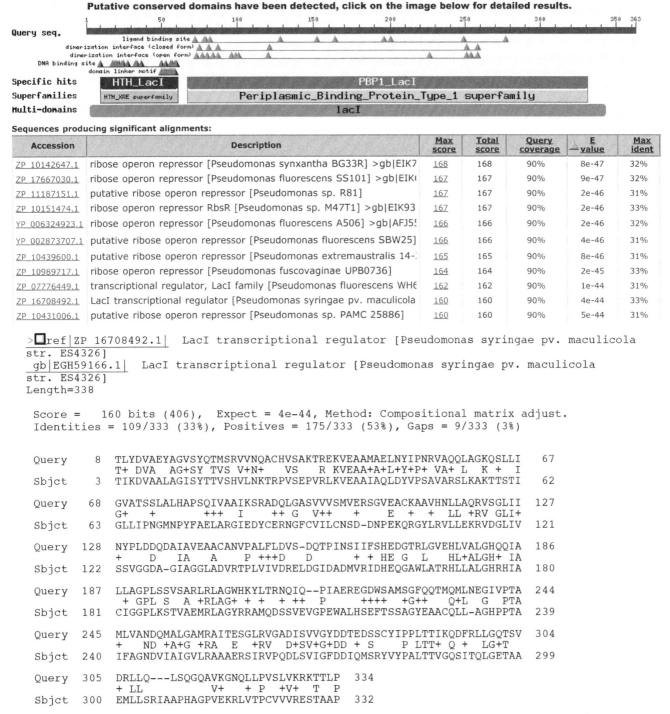

FIGURE c.2 **Comparison of the same protein from different species by aligning them with BLAST: the** *lac* **repressors of** *E. coli* **and** *Pseudomonas syringae.*

a portion of the sequence from the common ancestor in one of the sequences but not in the other. These gaps usually occur at locations in the tertiary structure of the properly folded polypeptide where a loop between secondary structure is found.

d. *Score.*

This increases as similarity in sequence increases. Several factors are used to determine this value including percent identities, percent positives, and the frequency and length of gaps. The methods used to calculate the score are summarized on the following NCBI Web page: http://www.ncbi.nlm.nih.gov/BLAST/tutorial/Altschul-1.html.

Do the results for the alignments that you made with BLAST or the alignment program of UniProt show a significant similarity between any two of these four sequences you compared?

What degree of similarity is necessary to suggest a common ancestry for two polypeptides?

To answer these questions it is necessary to determine a criterion for the minimum percentage of identity, percentage of positives, or alignment score that is indicative of common ancestry. First, there are only 20 types of amino acids found in proteins, so some similarity in sequence will occur by chance. Second, a stretch of 50% identities is much more significant over 100 amino acids than over 20.

There is no magic number for the percentage of identity, the percentage of positives, or the alignment score necessary to conclude that the sequences being compared demonstrate a common ancestry. The best way to get a feeling for proper data evaluation is to run alignments on some pairs of sequences that are very unlikely to have common ancestry and perform different functions. The polypeptides being evaluated in our exercises are 237–369 amino acids in length and have in common their ability to catalyze the oxidation or reduction of nicotinamide coenzymes. Retrieve 10 database entries for polypeptides in this size range that are unlikely to be related to these enzymes. Run the 10 pairwise *blastp* alignments or UniProt alignments as described above with one of the redox enzymes we are studying as query and the other polypeptides as subject. The mean and standard deviations of the parameters obtained for these 10 pairwise alignments should give you an idea of the values for the percentage of identity, the percentage of positives, or the alignment score that result when unrelated sequences of these lengths are compared. For convincing designations of relationship, the values of these parameters should be several standard deviations larger than these mean values for the alignments of the unrelated sequences.

It is also important that at least one of the comparisons provides a strong match so that it is clear that the program is being applied properly. The two bovine variants of LDH provide this positive control.

6. Make conclusions.

What may be concluded from this exercise? Does it appear that the domains for binding NAD(P) in FNR (amino acids 214–369), LuxG (amino acids 106–237), and LDH (amino acids 20–162 and 249–265 in the A isoenzymatic subunit, and 21–163 and 250–266 in the B isoenzymatic subunit) evolved from a common ancestor?

Is it possible for domains in two different proteins to have homologous tertiary structures that are descended from a common ancestor even if there is no similarity in sequence?

COMPUTATIONAL EXERCISE 2

Compare the Tertiary Structures of Polypeptides

The Macromolecular Modeling Database of the NCBI includes cartoons of the tertiary structures of the folded polypeptides in the crystallographic molecular models of enzymes as well as proteins not classified as enzymes. Initially, the arrangement of these cartoons in the database on this Web site may appear a bit ambiguous. From the menu on the Web site for the NCBI, choose *Structure* to gain access to the Macromolecular Modeling Database. The NCBI obtains the coordinates for the crystallographic molecular models registered in the Protein Data Bank of the Research Collaboratory for Structural Bioinformatics. Each identification number in the Protein Data Bank is four characters: numbers and uppercase letters. The same identification number is used on both Web sites. The Macromolecular Modeling Database also has its own identification system of five numerals. For this exercise, you should use the programs provided by the NCBI because they are easier to run than those provided by the Protein Data Bank.

These cartoons of the tertiary structures were drawn from the atomic coordinates of the actual tertiary structures of molecular models that were constructed from data obtained from the diffraction of X rays by crystals of these proteins, nuclear magnetic resonance spectroscopy of soluble molecules, or occasionally both methods used in combination. In some cases, the cartoon includes one or more small molecules that are bound to the protein in the crystal and appear in the crystallographic molecular model. Even though molecules of protein are usually oligomers of subunits, because of the way in which crystallography is performed, only one or two subunits of the larger oligomer may appear in the list of atomic coordinates for a particular protein in the Protein Data Bank. The entire quaternary structure of the protein can be obtained by applying symmetry operations to the coordinates that are listed, but this is not consequential to the present exercise.

The *Cn3D* software is an application that helps you with the Web browser. It animates images of the cartoons by rotating them. It is available on the NCBI site as a free download. The *Vector Alignment Search Tool* on this Web site is an application that assays for similarity in tertiary structure. This algorithm may discover homology among several proteins even if there is no significant similarity in their sequences of amino acids. Once a set of homologous structures is identified by the Vector Alignment Search Tool, *Cn3D* may be used to superpose the two structures (Kyte, 2007, pp 362–376). Examples of Cn3D images of cartoons of the tertiary structure of the *lac* repressor (Lewis et al., 1996) from the Macromolecular Modeling Database and a superposition by the Vector Alignment Search Tool of this structure upon a related protein from *Chromobacterium violaceum*, are shown (Figure c.3).

The Vector Alignment Search Tool will compare a single tertiary structure to all of the tertiary structures in the Macromolecular Modeling Database but, unlike BLAST, it is not designed to compare a small set of user-specified entries. There is an entry for FNR from *S. oleracea* in the Macromolecular Modeling Database, but as of July 2011 none for *P. leiognathi* LuxG or bovine LDH. Cartoons of tertiary structures of other closely related mammalian lactate dehydrogenases are in the Macromolecular Modeling Database. It also includes a cartoon of the tertiary structure of the enzyme (EC 1.5.1.41) riboflavin reductase [NAD(P)H] (Fre) encoded by the *fre* gene of *E. coli* that has 36% sequence identity to LuxG and also binds NAD (see Background: Flavin Reductases, Bioluminescence and the *lux* Operon). Since the common feature of the three enzymes

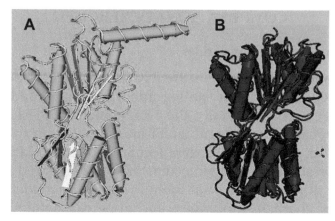

FIGURE c.3   Viewing and comparing polypeptide tertiary structures with *Cn3D* and *VAST*.   (A) A cartoon of the tertiary structure of a single *lac* repressor subunit from *E. coli* viewed with *Cn3D*. The crystallographic molecular model has the identification number 1LBI in the Protein Data Bank and the scientists that submitted the coordinates also published a summary (Lewis et al., 1996). The green cylinders with a cone at one end represent α helices and the flat, yellow arrows represent strands of β structure. The amino-terminal to carboxy-terminal orientation is shown by the cones and arrows. (B) A superposition of the cartoon of the tertiary structure of a subunit of *lac* repressor (magenta) upon a cartoon of the tertiary structure of the transcriptional regulator GntR from *Chromobacterium violaceum* (blue). The identification number for the atomic coordinates for the latter molecular model is 3H5O in the Protein Data Bank, and the coordinates were submitted by Z. Zhang, S.K. Burley, and S. Swaminathan.

is catalyzing the oxidation and reduction of NAD(P), this domain from one of the enzymes you will be studying will be used to search the entire MMDB. The goal is to determine if the domains for binding NAD(P) in the other enzymes catalyzing the oxidation and reduction of NAD(P) to be studied in the exercises in this manual (or their homologs from other species) have similar tertiary structures.

Procedure

1. Access the homepage for the Macromolecular Modeling Database at http://www.ncbi.nlm.nih.gov/sites/entrez?db=structure.

2. Click on *Resources* to get descriptions of the available programs.

3. Download and install *Cn3D* for your web browser.

4. Do a text search of the *Structure* database (using the enzyme and species names) for the appropriate entries and record the identification number from the Protein Data Bank of each.

5. Open the entry for the structure of *E. coli* Fre.

6. In the *Molecules and Interactions* box, click on *Show Sequence Annotation*.

7. A horizontal line diagram will appear, showing the two domains of this enzyme (Kyte, 2007, pp 376–391). The carboxy-terminal domain (right side, amino acids 97–232) is the domain for binding NAD(P).

8. Click on this part of the diagram to get scores for the comparison of the tertiary structure of Fre to the tertiary structures of various entries in the Macromolecular Modeling Database. There should be many pages of comparisons. As with the comparison of sequences with BLAST, it is important to determine the baseline score. Advance to the last page of the list to see the lowest score, then return to the top of the list. Those on the top of the list with significant scores are referred to as *Related Structures*.

9. Because structures of many NAD(P)-binding proteins are in the Macromolecular Modeling Database, the list will include many entries in addition to those relevant to this manual. Use the identification numbers from the Protein Data Bank obtained earlier to narrow the list to those of interest:

   a. Click *Advanced Related Structure Search.*

   b. In the *Find Related Structures* box, enter the appropriate identification numbers.

   c. Record the structure comparison scores on the new list.

10. If the scores appear significant, obtain an image of the superposition of the cartoons for the tertiary structures of these polypeptides: click on *View 3D Alignment with Cn3D.*

11. To get a color-coded depiction of secondary structure elements in the polypeptides: in *Global Style Options*, activate *Helix Objects* and *Strand Objects.*

Interpretation of Results

1. Does it appear that the enzymes you will study in the experimental exercises in this manual have similar tertiary structures in their domains for binding NAD(P)?

2. Consider these new results and those from the previous exercise. If there is similarity in tertiary structure, is there also similarity in the sequence of amino acids?

3. Using the references from the literature for the entries from the Macromolecular Modeling Database and PubMed (the NCBI literature database), find out what has been learned about the NAD(P)-binding domains of enzymes catalyzing the oxidation and reduction of NAD(P).

4. What is reasonable to conclude regarding the evolutionary time frame in which this domain arose in the lineages giving rise to extant bacteria, plants, and animals? Was it a single event? Can you place the time-point(s) relative to the branch-points on the diagram shown (Figure c.4)?

5. Do your results suggest that NAD(P) was an important redox metabolite in bacteria before eukaryotes evolved? Explain your reasoning.

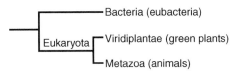

FIGURE c.4   **Branch-points during evolution.**

Purification and Characterization of Ferredoxin-NADP⁺ Reductase from Chloroplasts of *S. oleracea*

BACKGROUND: THE ROLE OF FERREDOXIN-NADP⁺ REDUCTASE IN LINEAR AND CYCLIC ELECTRON TRANSPORT IN PHOTOSYNTHESIS

The reactions that capture the energy of light during photosynthesis in plants and algae occur in a *thylakoid*. Thylakoids are convoluted membrane-enclosed structures tightly packed within the chloroplast (Figure 1.1). The large multisubunit complexes of photoreaction center II, plasto-quinol–plastocyanin reductase, and photoreaction center I in the linear electron transfer system of photosynthesis (Figure 1.2) are embedded within the membrane of a thylakoid (Barber, 2009; Lehninger et al., 2008; Ninfa et al., 2010; Shikanai, 2007).

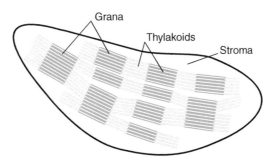

FIGURE 1.1   **Internal structure of a chloroplast in a plant cell.**   Each thylakoid is a membrane-enclosed compartment. Thylakoids are tightly packed together in a chloroplast. The interior of each thylakoid is known as its lumen. The stacked thylakoids are referred to as grana. The stroma is the portion of the chloroplast outside of the thylakoids.

The *photoreaction center* of *photosystem II* (PSII), which contains chlorophyll P680, uses the energy from four photons of light to oxidize two molecules of water to one molecule of oxygen. In each of the four steps of this photo-oxidation, an electron, ultimately from one of these molecules of water, reduces the oxidized ground state of chlorophyll P680, which is a strong enough oxidant to remove an electron from water. After its reduction and its photoexcitation, reduced chlorophyll P680 in the excited state has become a strong reductant and an electron in this excited state, now with a high reduction potential, participates in a series of downhill electron transfers, ultimately leading to the reduction of a molecule of plastocyanin. After the electron has departed from the excited state of chlorophyll P680 into this chain of electron carriers, the chlorophyll P680 has become oxidized. When this oxidized chlorophyll P680 returns to the ground state, it is again the strong oxidant it was before, ready and willing to remove another electron from water (Kyte, 1995, pp 101–106).

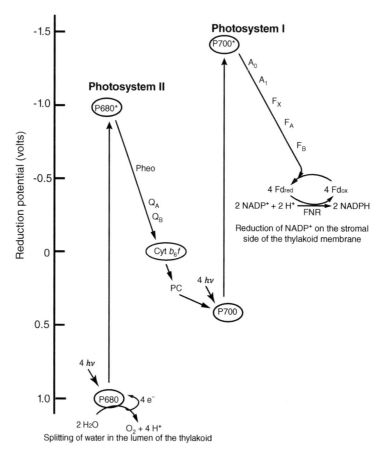

FIGURE 1.2   **Path of the electrons in the linear electron transfer system of photosynthesis.**   The abbreviations P680 and P700 are for the photoreaction centers of photosystem II and photosystem I respectively; Pheo is pheophytin, which is a molecule of chlorophyll lacking its central Mg^{2+}; Q_A and Q_B are plastoquinones A and B bound at different locations in one of the active sites of the photoreaction center of photosystem II; Cyt b_6f is plastoquinol–plastocyanin reductase; PC is plastocyanin, which is a soluble protein; A_0 is a particular chlorophyll; A_1 is a particular phylloquinone; F_X, F_A, and F_B are iron–sulfur clusters that are bound within the membrane; Fd is ferredoxin, which is a soluble protein. Ferredoxin-NADP⁺ reductase catalyzes the reduction of NADP⁺ that is the final step in the linear transfer of electrons accomplished by photosystem I in chloroplasts.

The second input of light energy occurs at the *photoreaction center* of *photosystem I* (PSI) when an additional photon is absorbed by chlorophyll P700 in photosystem I after it has been reduced by a plastocyanin. Upon photoexcitation, reduced P700 becomes a strong reductant in the excited state, and an electron in this excited state, now with a high reduction potential, participates in a series of downhill electron transfers, ultimately leading to the reduction of a molecule of ferredoxin.

To oxidize two molecules of water to one molecule of oxygen, four electrons must be removed, two from each molecule of water. This sets the stoichiometry of the overall reaction. Two molecules of water, four photons, and four molecules of oxidized ferredoxin are converted to one molecule of oxygen and four molecules of reduced ferredoxin. The four electrons from these four reduced ferredoxins are then used to reduce two molecules of NADP⁺ in a reaction catalyzed by ferredoxin-NADP⁺ reductase (FNR) (Bruns and Karplus, 1995; Karplus et al., 1991; Negi et al., 2008).

The NADPH generated by these photoreactions is used in the chloroplast and the cytoplasm for carbon fixation. Carbon fixation is the conversion of carbon dioxide to aldoses and ketoses. Although for historical reasons the reactions that accomplish carbon fixation are commonly referred to as the *dark reactions* of photosynthesis, they also occur when the plant is being exposed to light.

Ferredoxin-NADP⁺ reductase is encoded by a nuclear gene. A precursor 369 amino acids in length is synthesized on cytoplasmic ribosomes and then translocated into the chloroplast (Aliverti et al., 1990; Onda and Hase, 2004). A transit peptide is excised from the amino-terminal end of the original polypeptide during translocation to give the mature protein of 314 amino acids. The pattern of α helices and β strands in the tertiary structure of the mature protein has been observed experimentally by X-ray crystallography (Bruns and Karplus, 1995; Karplus et al., 1991) and is indicated on an alignment of the amino acid sequence of the protein with two related proteins (Figure 1.3B). As discussed earlier in the Introduction to Enzymes Catalyzing Oxidation–Reductions with the Coenzyme NAD(P), evidence available at this time suggests that the mature FNR may function in the chloroplast as both a monomer and dimer of its subunit (Lintala et al., 2007; Negi et al., 2008; Zanetti and Arosio, 1980).

In the enzymatic assay used in the following exercises, FNR oxidizes NADPH to NADP⁺. Electrons are transferred by the enzyme from NADPH to a dye and the resulting reduction of the dye produces a change in absorbance. The assay is convenient and inexpensive. The role of FNR in

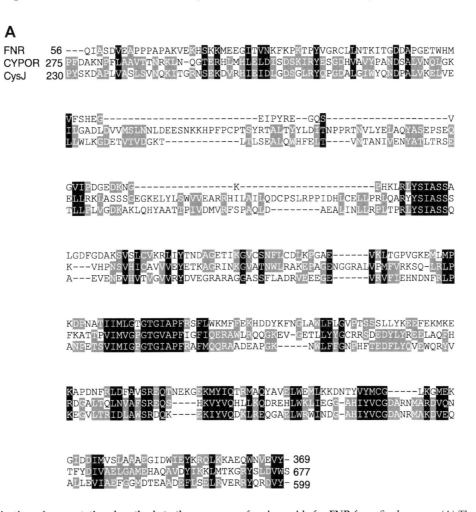

FIGURE 1.3 Application of computational methods to the sequence of amino acids for FNR from *S. oleracea*. (A) The primary structure of FNR from *Spinacia oleracea* can be aligned with that of human NADPH-hemoprotein reductase, labeled CYPOR (the same enzyme is also called P450 oxidoreductase and cytochrome P450 reductase), and sulfite reductase (NADPH), labeled CysJ, from *Escherichia coli*. The entire mature sequence of amino acids for FNR is aligned with the portions of NADPH-hemoprotein reductase and sulfite reductase (NADPH) that bind flavin adenine dinucleotide (FAD) and NADPH. The sequence upstream of these segments binds flavin mononucleotide (FMN), which is not a substrate or prosthetic group for FNR, but FNR does have FAD as a prosthetic group. Shading indicates similar or identical amino acids in three (darker) or two (lighter) of the sequences.

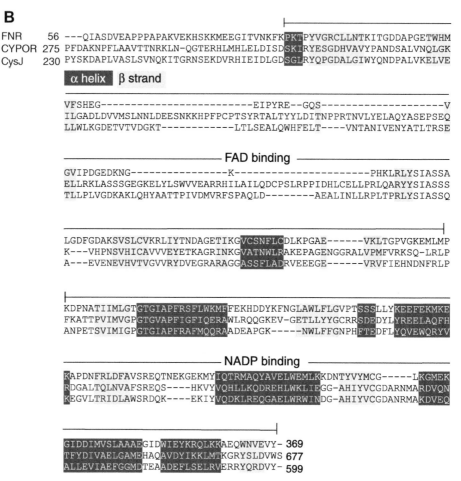

B

```
FNR      56 ---QIASDVEAPPPAPAKVEKHSKKMEEGITVNKFKPKTPYVGRCLLNTKITGDDAPGETWHM
CYPOR   275 PFDAKNPFLAAVTTNRKLN-QGTERHLMHLELDISDSKIRYESGDHVAVYPANDSALVNQLGK
CysJ    230 PYSKDAPLVASLSVNQKITGRNSEKDVRHIEIDLGDSGIRYQPGDALGIWYQNDPALVKELVE
```

α helix   β strand

```
VFSHEG--------------------------EIPYRE--GQS-------------------V
ILGADLDVVMSLNNLDEESNKKHPFPCPTSYRTALTYYLDITNPPRTNVLYELAQYASEPSEQ
LLWLKGDETVTVDGKT-----------LTLSEALQWHFELT----VNTANIVENYATLTRSE
```

——————————————————— FAD binding ———————————————————

```
GVIPDGEDKNG---------------K---------------------PHKLRLYSIASSA
ELLRKLASSSGEGKELYLSWVVEARRHILAILQDCPSLRPPIDHLCELLPRLQARYYSIASSS
TLLPLVGDKAKLQHYAATTPIVDMVRFSPAQLD--------AEALINLLRPLTPRLYSIASSQ
```

```
LGDFGDAKSVSLCVKRLIYTNDAGETIKGVCSNFLCDLKPGAE------VKLTGPVGKEMLMP
K---VHPNSVHICAVVVEYETKAGRINKGVATNWLRAKEPAGENGGRALVPMFVRKSQ-LRLP
A---EVENEVHVTVGVVRYDVEGRARAGGASSFLADRVEEEGE------VRVFIEHNDNFRLP
```

```
KDPNATIIMLGTGTGIAPFRSFLWKMFFEKHDDYKFNGLAWLFLGVPTSSSLLYKEEFEKMKE
FKATTPVIMVGPGTGVAPFIGFIQERAWLRQQGKEV-GETLLYYGCRRSDEDYLYREELAQFH
ANPETSVIMIGPGTGIAPFRAFMQQRAADEAPGK-----NWLFFGNPHFTEDFLYQVEWQRYV
```

——————————————————— NADP binding ———————————————————

```
KAPDNFRLDFAVSREQTNEKGEKMYIQTRMAQYAVELWEMLKKDNTYVYMCG-----LKGMEK
RDGALTQLNVAFSREQS----HKVYVQHLLKQDREHLWKLIEGG-AHIYVCGDARNMARDVQN
KEGVLTRIDLAWSRDQK----EKIYVQDKLREQGAELWRWINDG-AHIYVCGDANRMAKDVEQ
```

```
GIDDIMVSLAAAEGIDWIEYKRQLKKAEQWNVEVY- 369
TFYDIVAELGAMEHAQAVDYIKKLMTKGRYSLDVWS 677
ALLEVIAEFGGMDTEAADEFLSELRVERRYQRDVY- 599
```

FIGURE 1.3   **Cont'd** (B) The published tertiary structure of FNR is used to predict a pattern of α helices and β strands in the other two proteins. The alignment, identification of secondary structure, and shading were performed with the *Genedoc* software described in the computational section.

linear electron transport, however, which is the best characterized function of FNR in the chloroplast, is to reduce NADP⁺ to NADPH (Figures 1.2 and 1.4). This fact, however, is irrelevant when an assay for the enzyme is being designed. All enzymes catalyze their reactions in both directions so an enzyme can be assayed in either direction of its stoichiometric reaction, whichever is the

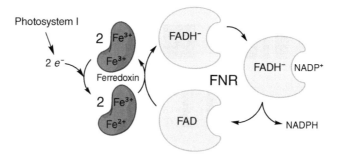

FIGURE 1.4   **Ferredoxin-NADP⁺ reductase catalyzes transfer of electrons from ferredoxin to NADP⁺ through its prosthetic FAD.** The route through which electrons are transferred from ferredoxin to NADP⁺ includes the prosthetic flavin adenine dinucleotide (FAD) (Figure i.2) tightly bound to the enzyme. This prosthetic group undergoes reduction and oxidation as the electrons pass through it. Each molecule of ferredoxin can provide one electron as one of its associated iron atoms changes from Fe^{2+} to Fe^{3+}. The sequential transfer of two electrons, one from each of two molecules of reduced ferredoxin that each in turn must dock with the enzyme and unload its electron, reduces FAD to FADH⁻. When the FADH⁻ then passes a hydride ion to NADP⁺, it is reduced to NADPH.

most convenient. As you will learn, the only purpose of an assay is to discover where the enzyme is located, and it makes no statement about the role of the enzyme in the cell. There are reports, however, that FNR also functions in cyclic electron transport (CET), which is a different route electrons take around photosystem I (Lintala et al., 2007; Zhang et al., 2001). If so, the role of the enzyme in this case is to oxidize NADPH to NADP$^+$, so the measurements of enzymatic assay will mimic the role of the enzyme in cyclic electron transport rather than its role in linear electron transport.

Searches of databases with the sequence of FNR reveal (Figure 1.3A) that it is closely related to two other proteins (Karplus et al., 1991): human NADPH-hemoprotein reductase (EC 1.6.2.4, labeled with the alternate name CYPOR) and the α subunit of sulfite reductase (NADPH) from the bacterium *Escherichia coli* (EC 1.8.1.2, labeled with the name of its gene, *CysJ*). More information about NADPH-hemoprotein reductase is in the literature (Gutierrez et al., 2003; Miller et al., 2009), and more information about sulfite reductase (NADPH) is also available in the literature (Kozmin et al., 2010; Zeghouf et al., 1998). Studies of the structure and function of FNR from *Spinacia oleracea* may provide insight into the mechanism of function of these other proteins and vice versa.

A procedure initially created for purification of H$^+$-transporting two-sector ATPase from chloroplasts (Binder et al., 1978) has been adapted for purification of FNR (Apley et al., 1985; Karplus et al., 1991). Although FNR may not share any of the properties of H$^+$-transporting two-sector ATPase, this method has proven to be a convenient way to isolate FNR.

FNR EXERCISE 1

Preparation of a Lysate of Chloroplasts and Sequestration of Proteins by a Solid Phase for Anion Exchange

Technical Background

The net charge on a molecule of protein is determined by the acid–bases it contains (Figure 1.5A) and the ions that are bound to it when it is dissolved in the solution being used for a particular experiment (Kyte, 2007, pp 32–36). The *isoionic point* of a protein is the pH of a solution containing only water and the protein, including all of its tightly bound ions, prosthetic groups, and post-translational modifications. In the range of neutral pH, pH 5 to 9, the concentration of protons and hydroxide ions is negligible relative to the concentration of charge on the protein, so the isoionic point of a protein is approximately the pH at which the protein, dissolved in pure water, has zero net charge.

The isoionic point of a protein can be estimated computationally by creating a theoretical titration curve of the protein. The numbers of each type of charged side chain from the amino acids and the prosthetic groups in the protein are tabulated based on the known sequence of amino acids for the protein and its content of prosthetic groups. The values of pK_a for the side chains in a fully

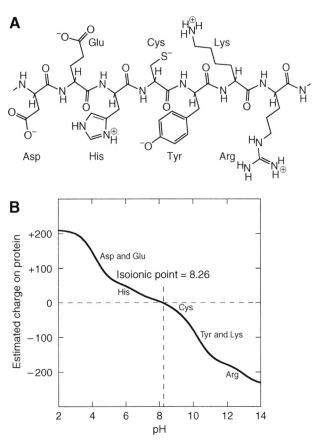

FIGURE 1.5 **The ionic properties of polypeptide chains.** (A) The ionic form of the side chain of each of the seven amino acids that contribute to the isoionic point of polypeptides. (B) Theoretical titration of the tetramer of fructose bisphosphate aldolase from *Oryctolagus cuniculus*. In a solution of very low pH, in the absence of any other ions, which is impossible in practice but possible in theory, the polypeptide has a net positive charge. As base is added and the pH increases, the positive charge on the polypeptides gradually decreases. At pH 8.26, the polypeptides have no net charge. As more base is added, the net charge on the polypeptides becomes negative. The regions in which each of the amino acid side chains in (A) titrate is indicated on the curve.

unfolded polypeptide (Kyte, 2007, pp 74–83) are aspartic acid (Asp or D) = 4.0, glutamic acid (Glu or E) = 4.3, histidine (His or H) = 6.6, cysteine (Cys or C) = 8.7, tyrosine (Tyr or Y) = 9.8, lysine (Lys or K) = 10.5, and arginine (Arg or R) = 13. A basic side chain (His, Lys, or Arg) has a net positive charge at a pH below its pK_a. An acidic side chain (Asp, Glu, Cys, or Tyr) has a net negative charge at a pH above its pK_a. From these values of pK_a, the numbers of each type of amino acid, and the values of pK_a for the prosthetic groups (for example, flavin adenine dinucleotide (FAD)), a theoretical titration curve for the protein, which is a plot of the estimated charge on the protein as a function of pH, is calculated. Such a curve can be calculated for fructose-bisphosphate aldolase from *Oryctolagus cuniculus* (Figure 1.5B). This analysis predicts that, when it is dissolved in water, the protein should have zero net charge at a pH of 8.3. Thus 8.3 is predicted to be the isoionic point of fructose-bisphosphate aldolase from *O. cuniculus*. The measured isoionic point of fructose-bisphosphate aldolase from *O. cuniculus* is 8.4 (Velick, 1949).

The *isoelectric point* of a protein (p*I*), however, is the pH at which, under the conditions of an experiment, the mean net molecular charge of the protein is zero. The isoelectric point of a protein is the number that is of interest when you are performing a particular procedure in which the protein is dissolved in a solution that contains ions from the buffer and the electrolytes. For example, the following experiment using ion exchange proceeds in solutions containing *N*-[1,3-dihydroxy-2-(hydroxymethyl)-2-propanyl]glycine (tricine), 2-amino-2-hydroxymethyl-propane-1,3-diol (Tris), sodium ion, and chloride ion. The isoelectric point of a protein is dependent on conditions and cannot be estimated as the isoionic point can be, because proteins bind anions and cations, such as sodium, Tris, and chloride, at sites that cannot be predicted. Consequently, these bound ions affect the charge unpredictably.

Experimental Plan

Spinach leaves are sheared, and their cells are disrupted in a blender. The reason that spinach leaves are used as a starting material is that they are available in supermarkets, they are inexpensive, and they have a reasonably high concentration of chloroplasts. An inhibitor of certain serine endopeptidases, phenylmethylsulfonyl fluoride, is included in the homogenization buffer mainly for historical reasons (Pringle, 1970). The intention is to inactivate serine endopeptidases that might digest FNR and inactivate its enzymatic activity.

Chloroplasts are isolated by four rounds of centrifugation and washing. After the washing, the chloroplasts are burst open by lowering the ionic strength of the preparation with a solution of 0.75 mM ethylenediaminetetraacetic acid (EDTA). The EDTA is included to chelate metal ions, some of which can inhibit enzymatic activity, and also inactivates metalloendopeptidases. Water flows rapidly into the chloroplasts because there is a higher concentration of molecules other than water, known as *osmolytes*, inside the chloroplast than outside. This rush of water into the chloroplasts causes them to burst.

Because FNR is translocated into the chloroplast, the amino-terminal 55 amino acids of the original polypeptide are removed to generate the *mature* polypeptide. Application of a program on the ExPASy database Web site (refer to Computational Techniques for Biochemistry) to the sequence of amino acids for mature FNR from *Spinacia oleracea* predicts that it will have an isoionic point of 6.3. The lysate of the chloroplasts is adjusted to a pH between 7.5 and 8.0 before the anion-exchange step. Although the isoelectric point of FNR in this solution cannot be predicted, the fact that the anion exchange you will be doing will be successful is strong evidence that the protein is anionic at these values of pH.

Diethylaminoethyl (DEAE)-methacrylate and DEAE-cellulose are both solid phases that have positive charges covalently attached to them, so they perform anion exchange (Figure 1.8) at values of pH below 10 where the diethylamino groups are protonated and positively charged (Kyte, 2007, pp 8–11 and 23–25). These fixed positively charged ammonio groups require that anions be present in the solution immediately adjacent to them to act as their counter ions. In the absence of other anions, the molecules of FNR are forced to fill this role of counter ion. Because they are forced to be the required counter ions, they are confined in the solution near the surface of the solid phase. As the concentration of other anions, in our case chloride ions, is raised, those other anions take over the role of counter ions for the fixed cationic charges of the DEAE groups because their concentration is so much greater than that of FNR, and FNR is released from the solution immediately adjacent to the fixed cationic charges into the outside solution. The concentration of chloride ions needed to release a protein from its role as a counter anion to a solid phase with fixed positive charge depends on the net negative charge on the protein, so each protein has a characteristic concentration of chloride at which it is released from the layer of solution immediately adjacent to the solid phase.

Although many other proteins from the chloroplasts are also anionic at this pH, this is a good way to initiate the isolation of FNR because it eliminates any cationic proteins and it concentrates the anionic proteins on the solid phase so that when they are released the solution that results has a high concentration of protein. All steps are performed at 4 °C or on ice to prevent denaturation of FNR.

Reagents, Supplies, and Equipment

~500 g of spinach leaves (stems removed).

DEAE-methacrylate or DEAE-cellulose in equilibration buffer (the volume of settled solid phase should be ~150 mL).

Equilibration buffer (1 L, in cold room):
  1 mM ATP.
  2 mM EDTA.
  50 mM Tris 7.8.

Homogenization buffer (1.5 L, in cold room):
  50 mM NaCl.
  5 mM $MgCl_2$.
  250 mM sorbitol.
  1 mM phenylmethylsulfonyl fluoride (toxic! Refer to Safety Guidelines for Biochemical Laboratories section).
  50 mM tricine, pH 8.

Chloroplast wash buffer (400 mL, in cold room):
  10 mM NaCl.
  2 mM tricine, pH 8.

Chloroplast lysis solution (2 L, in cold room):
  0.75 mM EDTA.

Acid to adjust pH:

1 M HCl (for lowering pH during lysis): need a maximum of 2.5 mL.

Materials:

Cheesecloth and large funnel for filtering homogenate.

Six centrifuge bottles, 250 mL capacity (Figure 1.6, be certain each screw cap has a rubber O-ring for proper seal).

2 L Erlenmeyer flask with stir bar.

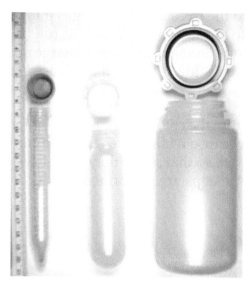

FIGURE 1.6 **Containers, with screw caps, for centrifugation.**   From left: 15 mL conical tube, Oak Ridge tube, and 250 mL centrifuge bottle. The cap for the centrifuge bottle has an O-ring that is critical to prevent leakage.

Equipment:

Stir plate.

Triple-beam balance.

Blender (Figure 1.7).

Beckman J2 preparative centrifuge with high-capacity JA-14 rotor (or equivalent), at 4 °C.

Precautions

Make certain you receive instructions for proper use of the preparative centrifuge (refer to Safety Guidelines for Biochemical Laboratories section). After balancing the sample bottles, be certain caps are tight to prevent leakage. Before initiating the spin, be certain the rotor cap is secure.

Do not use the centrifuge without supervision.

Procedure (One Laboratory Session)

The various stock solutions should be kept at 4 °C. When the preparations of chloroplasts and proteins are not in the cold room or the centrifuge, keep them on ice as much as possible.

Release of chloroplasts from the cells in the leaves (20 min).

FIGURE 1.7 **A kitchen blender.** Used to disrupt tissues and release cells.

You will be provided with a batch of spinach leaves. Determine and record the exact mass with a triple-beam balance.

In the cold room, use a kitchen blender (Figure 1.7) to fragment the leaves using the homogenization buffer (1 L total, this may be done in multiple batches).

Remove the coarse insoluble material by filtration through cheesecloth.

Transfer the filtrate to an appropriate number of centrifuge bottles (Figure 1.6).

Pelleting and washing of the chloroplasts (2 h).

Spin down chloroplasts and other cellular fragments at 8000 rpm (10,000 × g) for 15 min.

Decant and discard the supernatant.

Resuspend the pellets in a total of 500 mL of fresh homogenization buffer.

The pellet should be dispersed with a plastic pipette (vortexing not necessary).

Spin as before and decant the supernatant.

Resuspend the pellets in a total of 250 mL of chloroplast wash buffer.

Spin at 13,000 rpm (26,000 × g) for 30 min. Decant the supernatant.

Resuspend the pellets in a total of 125 mL of chloroplast wash buffer.

Spin at 13,000 rpm (26,000 × g) for 30 min. Decant the supernatant. Record the appearance of the pellet in your notebook.

Lysis of chloroplasts and binding of proteins to DEAE solid phase (1.5 h followed by an overnight step).

- Resuspend, in their bottles, each of the pellets containing the chloroplasts in ~100 mL of chloroplast lysis solution.

- Transfer all of the lysates to the 2-L Erlenmeyer flask (with stir bar) in the cold room and start stirring the suspension.

- Add more chloroplast lysis solution to a final volume of 2 L.

- Check the pH of the stirring lysate with indicator paper. It should be basic. Record the pH.

- Bring the pH down to 7.5–8.0 by adding HCl. Record the volume of acid added. You must have the suspension stirring while you adjust the pH to avoid local regions of very low pH around the drops of acid that you are adding that would denature the proteins.

- Allow the lysate to stir for 60 min then add the DEAE solid phase. Remember to check the speed of the stirrer as it warms up.

- Leave preparation stirring gently on the stir plate until the next laboratory session. Do not permit any foaming to occur.

Analysis of the Data and Questions to Answer in Your Report

1. Record the volume of each of the solutions that contained the chloroplast before and after lysis.

2. Why are the chloroplasts and other fragments segregated from the remainder of the broken cells including their cytoplasm before lysing them and adding the anion-exchange solid phase?

3. In this exercise, the cells in the leaves of spinach are shattered by the shearing forces created by the spinning blades of the blender, but in the exercises for the purification of LuxG-hexahistidine described later in this book, a combination of an enzyme (lysozyme), which hydrolyzes the outer membranes of the cells, and high-frequency sonication, which creates more severe shearing forces of lower amplitude, is used to lyse the *E. coli* bacteria. Why is a greater shearing force needed to lyse bacteria than to lyse plant cells? To answer this question, you should consult the textbooks that describe the structure of these cell types (Lehninger et al., 2008; Ninfa et al., 2010; Voet and Voet, 2010).

4. After lysing the cells in the leaves with the shearing force of the blender, a centrifugation at 10,000 × *g* is used to pellet the chloroplasts and other cellular debris. The supernatant is discarded. Again, consult textbooks on the subject (Boyer, 2012; Ninfa et al., 2010) to determine if the chloroplast fraction will also contain nuclei. Provide an explanation for your answer.

5. The chloroplast is surrounded by a double membrane. Both the inner and outer membranes consist of a lipid bilayer. The interior of a lipid bilayer consists of alkyl chains and is thus nonpolar. The osmotic shock method used to lyse the chloroplasts requires that water molecules cross these bilayers. Water is a polar molecule. Consult textbooks (Berg et al., 2012; Voet and Voet, 2010) and discuss how water molecules can enter chloroplasts in the procedure you performed. *Hint*: There is one mechanism that works for both the inner and outer membranes and one that works for only one of these membranes.

FNR EXERCISE 2

Elution of Proteins from the Solid Phase for Anion Exchange and Assays for the Concentration of Total Protein and FNR Activity

Technical Background and Experimental Plan

The DEAE solid phase used to capture FNR was prepared in equilibration buffer before being used in the previous exercise. This solution includes 1 mM ATP. There was no explanation of why ATP was used by the research group of Professor Engelbrecht to purify FNR (Apley et al., 1985). The main anions in the equilibration buffer, however, are chloride anions (approximately 25 mM of free chloride anion in the 150 mL), and the anions that are the counter ions to the DEAE groups on the solid phase before you add it to the solution containing FNR are chloride anions (approximately 150 mM chloride fixed as counter ions). When the 150 mL of solid phase is diluted into the 2 L of 0.75 mM EDTA, which contains almost no anions, the anionic molecules of protein become the main anions in the final suspension. In addition, because the ability of an anion to function as a counter ion in the solution surrounding a layer of fixed monocations is proportional to its molar concentration raised to the power of its net charge (Kyte, 2007, pp 8–11; Stein, 1967, pp 60–62), the polyanionic molecules of protein (Figure 1.5B) are much more effective counter ions for the DEAE groups than monoanionic chloride ions. For both of these reasons, the molecules of anionic protein replace most of the chloride ions as counter ions held in the solution immediately adjacent to the DEAE groups on the solid phase (Figure 1.8A).

The greater the net anionic charge on a particular molecule of protein, the higher the concentration of monoanionic chloride that must be used to displace it from the layer of solution surrounding the DEAE groups, so each molecule of protein is released at a characteristic concentration of chloride ion (Figure 1.8B). In the procedure you will use, the concentration of chloride ion will be increased in steps, and FNR will not be released from the solid phase until the concentration of chloride exceeds the concentration of chloride ions that can elute it specifically.

As in the previous exercise, most of the procedure is performed in the cold room. The only exception is the collection of the protein bound to the DEAE solid phase in the Buchner funnel; this may be at room temperature if no vacuum line is available in a cold room.

The material eluted from the solid phase in each of these steps in the concentration of chloride ion is assayed for activity of FNR so that fractions containing the majority of the enzyme can be pooled. In the assay you will be using, the reaction catalyzed by the enzyme is the oxidation of NADPH to NADP$^+$ (Figure 1.9). The NADPH reduces the prosthetic FAD on the enzyme and the prosthetic FADH$_2$ then reduces the chromogen, 2-(4-iodophenyl)-3-(4-nitrophenyl)-5-phenyl-2H-tetrazolium chloride (also known vernacularly as iodonitrotetrazolium or INT) (Cross et al., 1994). The formazan generated when INT is reduced has a strong absorption with a maximum at 490 nm. Measurement of the rate of increase in A_{490} (ΔA_{490} min^{-1}) allows for calculation of the concentration of enzyme activity. The rate of increase in absorbance is converted to standard units for the observed concentration of enzymatic activity (μmol min^{-1} mL^{-1}) using the Beer–Lambert Law and the *extinction coefficient* for the formazan at 490 nm (ε_{490}) of 18.5 mM^{-1} cm^{-1} (Zanetti, 1981). The unit of cm^{-1} in the conversion is the length of the cuvette or the length of the path that the light takes through the solution used in the spectrophotometer or the plate reader, and absorbance is a unitless quantity. The observed concentration of enzymatic activity in these units is also known as the initial velocity (v) of the enzymatic reaction.

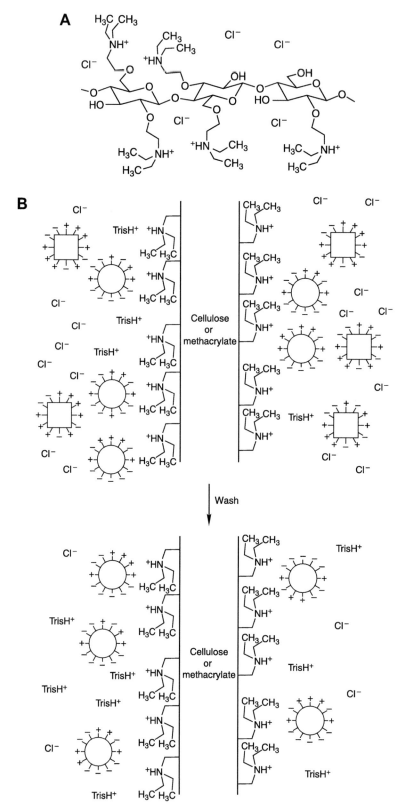

FIGURE 1.8 **Selection of proteins from a complex mixture by anion-exchange.**   (A) The structure of DEAE-cellulose. Cellulose is a polymer of glucose; DEAE groups are attached through ether linkages. In the pH range (7.5–8.0) used for binding of proteins from the chloroplasts followed by their elution, the tertiary amine of DEAE has a positive charge due to its protonation. The DEAE groups are coupled at random to the hydroxy groups of the cellulose. The counter ions in our case are chloride ions that are free and mobile in the solution adjacent to the DEAE. (B) A lysate of cells or chloroplasts contains many proteins. If the pH of the lysate is 7.5–8.0, some proteins will have a net negative charge (anionic, circles) and some a net positive charge (cationic, squares). If the solutions used for preparing the lysate and the DEAE solid phase have a low

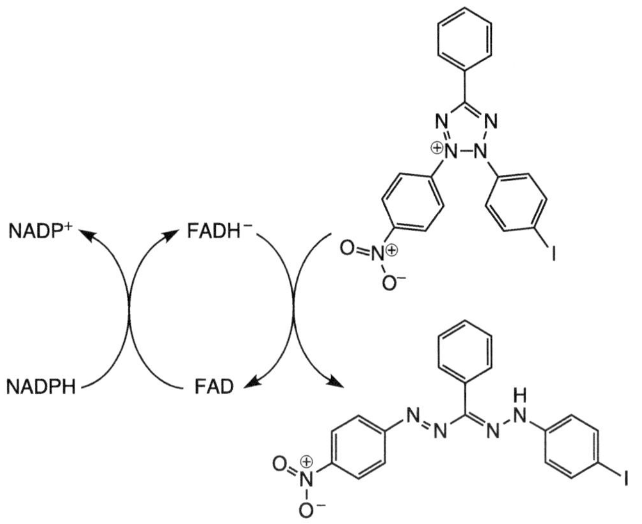

FIGURE 1.9   A chromogenic compound is used to detect the enzymatic activity of FNR. The oxidized form of INT is chromogenic. A chromogenic compound is initially colorless but turns colored upon oxidation or reduction. When NADPH is oxidized, a hydride ion is transferred to the isoalloxazine of the prosthetic FAD on the enzyme. The FADH⁻ then transfers two electrons to the tetrazolium of the oxidized chromogen and loses a proton to the solution. The transfer of the two electrons opens the tetrazolium and in the process the product picks up one proton from the solution that neutralizes the negative charge contributed by the electrons. This series of steps creates a *formazan* from INT. The formazan formed from INT has a violet color with maximum absorbance at 490 nm.

The concentration of total protein in the resulting pools is then estimated with the Bradford assay (Boyer, 2012; Bradford, 1976). In this technique, the binding of Coomassie Brilliant Blue G-250 (Figure 1.10) to molecules of protein results in a shift of its maximum of absorption from 465 to 595 nm. Although every type of protein binds a different amount of Coomassie Brilliant Blue for

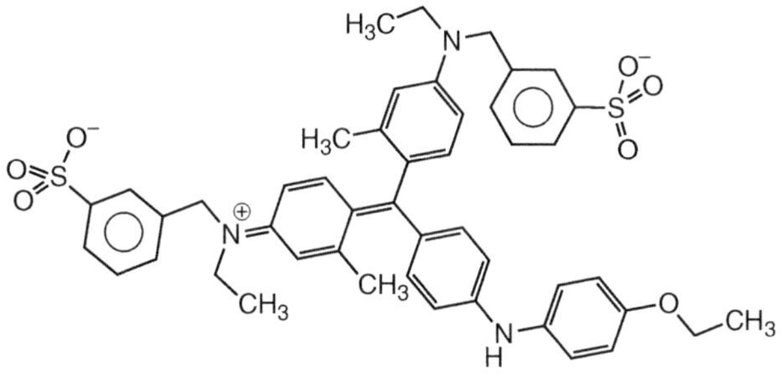

FIGURE 1.10   Coomassie Brilliant Blue G-250.

each gram of protein, causing the assay to be unquantitative, the differences in binding are not great and the assay is so easy that it is used when the protein is not pure or when large numbers of samples are to be assayed. If precise concentrations of a protein are needed, other assays are used (Kyte, 2007, p. 21; Moczydlowski and Fortes, 1981), but none has high accuracy. A standard curve is established for the Bradford assay by measuring the A_{595} of samples of different concentrations of γ-globulin (IgG heavy chain), a reference protein, mixed with the stain. The concentration of

concentration of anions, the polyanionic molecules of protein replace the monoanions associated with the DEAE groups as counter ions in the solution surrounding the fixed DEAE cations on the solid phase. Mobile ions of TrisH⁺ and a few chloride ions fill in to make the local regions of the solution electroneutral. Cationic proteins are washed away in an excess of buffer with a low ionic strength. Anionic proteins are then released from the solution surrounding the DEAE groups on the solid phase when a solution with a high enough concentration of monoanions, in our case chloride ions, is applied.

protein in the pooled fractions is then estimated by interpolation from this curve. The concentration of enzymatic activity (μmol min^{-1} mL^{-1}) and the estimates of the concentration of protein (mg^{-1} mL^{-1}) may then be used to calculate the observed *specific activity* of the enzyme preparation (μmol min^{-1} (mg of protein)$^{-1}$). This value is a measure of purity that is typically calculated at each stage of the purification of an enzyme.

Reagents, Supplies, and Equipment

Concentrated sodium chloride:
   5 M NaCl (need ~9 mL).

Buffer of intermediate anionic concentration for elution from DEAE (150 mL):
   Add concentrated NaCl to a portion of the equilibration buffer (see previous exercise) to give 100 mM final concentration of NaCl.

Buffer of high anionic concentration for elution from DEAE (150 mL):
   Add concentrated NaCl to a portion of the equilibration buffer to give 200 mM final concentration of NaCl. (Recall that each mL of 5 M NaCl contains 5 mmoles of NaCl.)

Reagents for assay of enzymatic activity (four stock solutions):
1. 5 mM NADPH in 50 mM sodium bicarbonate buffer, pH 9.
2. 5 mM INT in 70% ethanol, 5% Triton X-100.
3. 1 M TrisCl, pH 9.
4. 5 M NaCl.

Reagents for the Bradford assay:
1. Solution of standard protein: bovine γ-globulin (IgG), 0.1 or 1 mg mL^{-1} (stored at 4 °C).

2. Staining reagent (one mixture, prepared by laboratory staff, stored at room temperature): 0.01% (m/v) Coomassie Brilliant Blue G-250, 4.8% (v/v) ethanol, 8.5% (v/v) phosphoric acid.

Miscellaneous supplies:
   Large (~10-cm dia.) Buchner funnel and Whatman filter paper disk.

   Two 5-L sidearm-Erlenmeyer flasks.

   Glass chromatographic column with stopcock with a diameter of 5 cm and a height of 20 cm, mounted on rack in cold room.

   Twelve conical tissue-culture tubes (50 mL, with screw cap) for collection of the fractions from the steps in concentration of salt.

Spectrophotometric supplies and equipment:
   If you are using an automated plate reader (0.2-mL sample):
      Standard plastic multiwell plate (appropriate for wavelengths ≥ 400 nm; Figure 1.13 and Appendix I).

   If you are using a conventional spectrophotometer:

Standard plastic disposable cuvettes (appropriate for wavelengths ≥ 400 nm) with 0.5-mL chamber.

If you are using a Spectronics 20 D+ (Spec 20):
Glass culture tubes transparent to wavelengths ≥ 340 nm are used, each with a volume of 3 mL.

Automated plate reader (Figure 1.13 and Appendix I), conventional spectrophotometer, or Spec 20.

Miscellaneous equipment:
Vacuum port and tubing for sidearm-Erlenmeyer flask (in cold room if possible, ambient temperature is, however, acceptable).

Precaution

The Bradford reagent contains phosphoric acid that can erode flesh and Coomassie Brilliant Blue G-250 that is toxic. Wear gloves.

Procedure (Two Laboratory Sessions)

PART 1. WASH THE SOLID PHASE TO WHICH THE PROTEINS ARE BOUND AND THEN ELUTE THE PROTEINS

Collect the solid phase and transfer it to a column for zone elution (40 min).

Check the pH of the suspension of DEAE solid phase and the lysate of the chloroplasts with indicator paper and record the pH in your notebook. It should be close to the value recorded before addition of the DEAE solid phase.

Use the Buchner funnel and vacuum system to collect the solid phase on filter paper (Figure 1.11). It is best to perform this step at 4 °C. If no vacuum line is available in the cold room, however, working at ambient temperature is acceptable.

Wash the solid phase in the funnel with 200 mL of equilibration buffer.

Scrape the solid phase into a beaker with a spatula, and make a slurry by adding 150 mL of equilibration buffer.

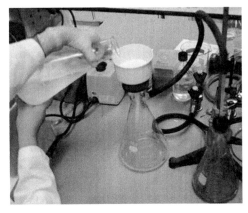

FIGURE 1.11 **Recovery of the solid phase with which the protein is associated by filtration.** After association of the anionic protein with an anion-exchange solid phase, a Buchner funnel supporting a circle of filter paper and inserted into a vacuum flask is used to collect the solid phase with which the protein is associated.

- Be certain the valve on the chromatographic column is closed.

- In the cold room: pour the slurry of solid phase into the column and allow it to settle (Figure 1.12A).

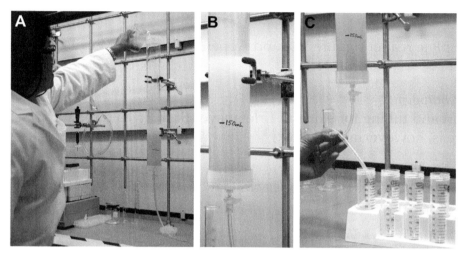

FIGURE 1.12 **Wash and elution of the anion-exchange solid phase.** (A) Pouring anion-exchange solid phase with bound protein into the column for wash and elution. After mixing the solid phase with the appropriate solution, the slurry is poured into a chromatographic column. (B) Waiting for the solid phase to settle. The *bed volume* may be determined by calculating the volume from the dimensions of the column and the height of the bed after the solid phase has packed into the column. (C) Manually collecting fractions of the eluate from the anion-exchange solid phase. If only a small number of fractions for each anionic concentration need be collected, an automated fraction collector is not necessary.

- Calculate the volume of packed solid phase (the *bed volume*, Figure 1.12B).

Wash the solid phase with a low concentration of chloride ion (15 min).

- Add 200 mL of equilibration buffer to the column and open the valve to allow it to flow through the solid phase (washing away unbound protein). Discard this wash.

Wash solid phase at the intermediate concentration (100 mM) of chloride ion (30 min).

- Add 150 mL of the buffer of intermediate anionic concentration, and collect six fractions of 25 mL (Figure 1.12C).

Wash solid phase at the high concentration (200 mM) of chloride ion (30 min).

- Add 150 mL of the buffer of high anionic concentration, and collect six fractions of 25 mL.

Store the 12 fractions in the cold room until you are prepared to assay them for enzymatic activity and the concentration of protein.

PART 2. ASSAY FOR THE ENZYMATIC ACTIVITY AND THE CONCENTRATION OF PROTEIN IN THE FRACTIONS ELUTED FROM THE SOLID PHASE

Assay for the activity of FNR in fractions eluted from the solid phase (2 h).

• Using the stock solutions, prepare a series of enzymatic reactions so that the following final concentrations will be achieved after you have added a sample from a fraction:

0.2 mM NADPH
0.5 mM INT
70 mM NaCl
200 mM Tris, pH 9

Add an appropriate volume from each of the fractions you have collected in turn. Begin taking A_{490} readings as quickly as possible after the sample from the fraction has been added. Assay a sample from each of the 12 fractions. Take measurements every 30 s for 3 min.

The following volumes are suggested for assaying the fractions:
Plate reader: sample of 10 μL from a fraction in a final volume of 0.2 mL.
Conventional spectrophotometer: sample of 25 μL from a fraction in a final volume of 0.5 mL.
Spec 20*: sample of 150 μL from a fraction in a final volume of 3 mL.

(Instructions for the use of multiwell plates and the automated plate reader (Figure 1.13) and instructions for the Spec 20, are in Appendices I and II respectively. If you are using a multiwell plate, all samples must be the same volume because the path length is proportional to this volume.)

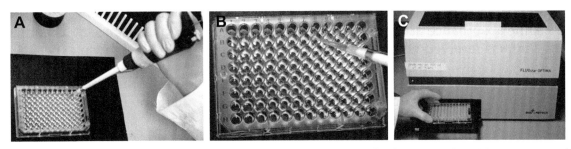

FIGURE 1.13 **Using a multiwell plate and an automated plate reader for assays of either enzymatic activity or concentration of protein.** (A and B) Loading samples into plate. A 96-well plate is shown: 8 rows (A–H) and 12 columns (1–12). A typical sample volume for these plates is 200 μL. For quantification of the absorption of visible light, a plate made of conventional plastic is used. For measurement of absorption in the ultraviolet range, a plate made from a plastic that is transparent to ultraviolet light is necessary. (C) After samples are pipetted into the wells, the plate is loaded into the reader for scanning.

When preparing identical reaction mixtures for assay of several samples of equal volume from the fractions, it is usually best to prepare a mixture of the reagents of sufficient volume for all reactions. Appropriate volumes of this mixture are pipetted into each well, cuvette, or tube. The sample from each fraction is then added to one of these wells, to the cuvette, or to the tube, which is agitated to mix the solution, and the change of absorbance at 490 nm is then monitored. The assay is then performed on the next fraction, and so forth.

The assay for concentration of enzymatic activity is most reliable if the rate of change in the absorbance at 490 nm, dA_{490}/dt, that you will be recording is between 0.05 and 0.2 min^{-1}. If the rate is not in this range, repeat the assays with an appropriate adjustment of the volume of that particular sample from the fraction, unless there is negligible activity in the fraction. Fractions with negligible activity are usually those that are collected later in each zone elution.

- Pool the fractions containing the majority of the FNR activity and record the final volume. If a significant amount of FNR activity is detected in both the intermediate and high anionic elutions, prepare separate pools of the fractions from each. Each of these pools will be referred to as a *pool from DEAE*, and if there are more than one of them, they should be numbered. To identify peak fractions, only the relative but not the absolute activities need to be compared. Do not pool all of the fractions containing any activity, because that will dilute the enzyme too much. Pool just the fractions with the majority of the activity. This is a decision you must make for yourselves. Calculation of the absolute activity in standard units is done after the fractions have been pooled.

- Finally, perform an accurate assay for enzymatic activity on a sample from each pool.

- Store each pool in the cold room.

Estimating the concentration of protein in a pool from DEAE by the Bradford assay (1.5 h).

The three types of spectrophotometers and sample containers described for the activity assay can also be used for the Bradford assay to estimate the concentration of protein. Decide which of the three instruments to use, then plan the reactions so that each receives the same volume of staining reagent. After adding the sample of a certain volume from the pool(s), add appropriate volumes of water to bring each assay to the appropriate final volume. Allow 20 min at room temperature for the color to develop fully before measuring absorbance. Although the peak absorbance of unbound Coomassie Brilliant Blue G-250 is at 465 nm, the dye has some absorbance at 595 nm. It is critical to subtract the absorbance at 595 nm, A_{595}, of an assay that has received no protein from all of the values of A_{595} in a given experiment.

Although this is not an issue for this preparation, you should know that this assay is not reliable for samples containing detergents such as dodecyl sulfate or Triton X-100.

- *Standard curve.* For the analysis of the data it is easiest to plot A_{595} as a function of the mass of γ-globulin in each reaction. After determining what the volume of the assay will be, calculate the masses of γ-globulin needed to prepare seven reactions with the following final concentrations: 0, 5, 10, 15, 20, 25, and 30 μg mL^{-1}. These masses that you have just calculated and that you will be adding to the well, the cuvette, or the tube will be used for plotting of the data, not the final concentrations in the assays. Prepare these seven reactions and assay the staining intensity as described above.

Samples from the pools from DEAE for Bradford assay.

If you are using a plate reader with a final volume for the assay of 0.2 mL, try 5, 10, or 20 μL.

If you are using a conventional spectrophotometer with a final volume for the assay of 0.5 mL, try 13, 25, or 50 μL.

If you are using a Spec 20 with a final volume for the assay of 3 mL, try 75, 150, or 300 μL.

- If any of the absorbances from the three samples are outside the range of the standard curve, assay a different volume as appropriate.
- Store the pools from DEAE in the cold room until you are ready for the next step in the purification.

Analysis of the Data and the Decision You Have to Make

1. Determining the concentration of protein in the pools by the Bradford assay:
   a. *Standard curve.* Plot A_{595} as a function of the mass of γ-globulin you added to each assay.
   b. *Interpolation.* Using this standard curve, determine, by interpolation from the A_{595} of each assay, the mass of protein in each sample from the pools that you assayed.
   c. *Quantify the total protein in each pool from DEAE.* From the masses of protein in each sample and the volume of the sample, calculate first the concentration of the protein in milligrams milliliter^{-1} in each pool and then the total mass of protein in each pool.

2. Calculating concentration of enzymatic activity from dA_{490}/dt:

   It is easy to plot kinetic data manually on graph paper. Plot A_{490} as a function of time. Draw the best straight line through the points. You can plot all of the lines on the same sheet of graph paper. You should use your judgment in drawing each line, disregarding a single point, for example, that doesn't conform to the others. The term *initial* velocity (v) emphasizes that you want the rate of the enzymatic reaction before it starts to decrease as reactant is consumed and equilibrium is established, so also disregard any points that seem to be falling off the line at later times.

   The initial velocity is equivalent to the concentration of enzymatic activity. In the exercise addressing the Michaelis constant of FNR from its steady-state kinetics, this quantity will be consistently called the initial velocity. For the calculations that will be performed below, however, the term "concentration of enzymatic activity" is more appropriate.

   If you prefer to use software to fit a line to the data by least-squares analysis, refer to the computational section for suggestions on sources. Remember, however, that a program does not know that the later points may deviate from linearity nor can it decide to ignore an obviously incorrect point. If you use software to fit the data, you must always look at each fit in graphical form to convince yourself that the line fit by the software is reasonable. Having to check the fit and having to correct the errors made by the computer makes fitting by computer more time consuming in this instance. Remember, you are always more intelligent than a computer.

   The rate dA_{490}/dt (min^{-1}) is the slope of this line. Remembering that micromoles milliliter^{-1} is the same as millimoles liter^{-1}, which is mM, use the extinction coefficient ε_{490} (18.5 mM^{-1} cm^{-1}) for the formazan of INT to convert dA_{490}/dt (min^{-1}) to $d[\text{formazan}]/dt$ (μmol mL^{-1} min^{-1}). The standard units for concentration of enzymatic activity are micromoles milliliter^{-1} minute^{-1}. One must remember, however, that the original sample from each pool was diluted when added to the assay. To obtain the concentration of enzymatic activity in the original undiluted sample of enzyme being characterized, correct the concentration of enzymatic activity in the assay for the dilution that occurred.

   The path-length for the cuvettes used in a conventional spectrophotometer is exactly 1 cm. The culture tubes used in a Spec 20 are considered to be 1 cm even though, because they are round, they are not. For samples in a multiwell plate, the path-length is < 1 cm. Most plate readers, however, have control programs that will correct the measured absorbance to correspond to a 1 cm path. Thus, path-length is usually 1 cm.

For those of you who enjoy equations rather than simply paying attention to units, the Beer–Lambert law is:

$$A_{490}(\varepsilon_{490}\, l)^{-1} = c \tag{1.1}$$

where l is the path-length in cm and c is the concentration of the absorbing molecule in mM. Even if you enjoy equations, you must make sure that the units cancel properly and that the number at the end has units attached to it.

Again, if you enjoy equations:

$$\frac{d\,[\text{formazan}]}{dt} \times \frac{(\text{reaction volume in } \mu L)}{(\text{sample volume in } \mu L)} = \text{concentration of FNR activity in the pool from DEAE} \tag{1.2}$$

Remember that for each mole of formazan produced, a mole of NADPH is oxidized, so the final formal units for concentration of enzymatic activity are micromoles NADP⁺ milliliter^{-1} minute^{-1}.

3. Use the concentration of activity and the volume of each pool to calculate the total enzymatic activity (μmol min^{-1}) in the pool.

4. Determine the specific activity of the pools from DEAE:

The *specific activity* is the amount of enzymatic activity for every unit mass of total protein. If the preparation is a mixture of proteins, the mass of protein that you estimated to be in each pool includes all the proteins, not just the enzyme the activity of which is being quantified. Specific activity is typically expressed as the number of micromoles of reactant (NADPH in this case) converted to product (NADP⁺ in this case), every minute for every milligram of protein (μmol min^{-1} mg^{-1}).

$$\frac{\left(\text{concentration of activity in } \mu\text{mol min}^{-1}\,\text{mL}^{-1}\right)}{\left(\text{concentration of protein in mg mL}^{-1}\right)} = \text{specific activity in } \mu\text{mol min}^{-1}\,\text{mg}^{-1} \tag{1.3}$$

During a purification, the specific activity usually increases because proteins other than the relevant enzyme are gradually being eliminated. The specific activity reaches a maximum level when the enzyme is pure. Consequently, when you are purifying a protein, you must decide between yield, the amount of total enzymatic activity, and purity, which is measured by the specific activity of the preparation. If any of the pools has a much lower yield of enzymatic activity and a low specific activity, it is not worth including it in the pool you will take on to the next step. You do not, however, want to throw away a significant amount of the enzyme just because the specific activity of a pool with a lot of enzyme happens to be lower than that of another pool. If you have saved two pools, decide whether to take both or only one on to the next step. If you decide to take both, mix them, and use the numbers you have already calculated for the two pools to calculate the specific activity and total activity of the FNR in the final pool.

Questions to Answer about This Exercise

1. Did you notice any color in any of the fractions eluted from the anion exchange? Suggest an explanation for your observations.

2. Diethylaminoethyl-cellulose and DEAE-methacrylate are anion exchangers. Carboxymethyl-sepharose is a cation exchanger (Boyer, 2012; Ninfa et al., 2010). If carboxymethyl-sepharose were used for enrichment of FNR from *S. oleracea*, how would the procedure have to be modified? Discuss whether or not such a modification might affect the activity of the enzyme.

3. You used anion exchange to collect the FNR from the chloroplast lysate. Discuss whether or not this technique also purified FNR.

4. The Bradford staining solution has color even before protein is added. The solution absorbs at both 465 and 595 nm. What are two critical features of this staining method that one should know to use it properly?

FNR EXERCISE 3

Isolation of FNR by Adsorption and Elution

Technical Background

This step in the procedure for purifying FNR is based on the observation that many proteins have a strong adventitious affinity for certain textile dyes (Boyer, 2012; Kyte, 2007, pp 25–26). If the dye is attached to a solid phase, the protein will adsorb to it. In this instance, unlike anion exchange, the protein actually binds to the dye on the solid phase, becoming noncovalently attached to the dye and hence the solid phase, which is covalently attached to the dye. In some cases, the specificity of the association between protein and dye is high enough that the majority of the unwanted proteins, which do not have the same affinity, are eliminated. The research group of Professor Engelbrecht (Apley et al., 1985), during their development of the procedure for purifying FNR, found that the enzyme has a high affinity for two such dyes, Procion Red and Cibacron Blue (Figure 1.14).You will use either Procion-Red-cellulose or Cibacron-Blue-agarose in this exercise to adsorb FNR from the pool from DEAE (Figure 1.15).

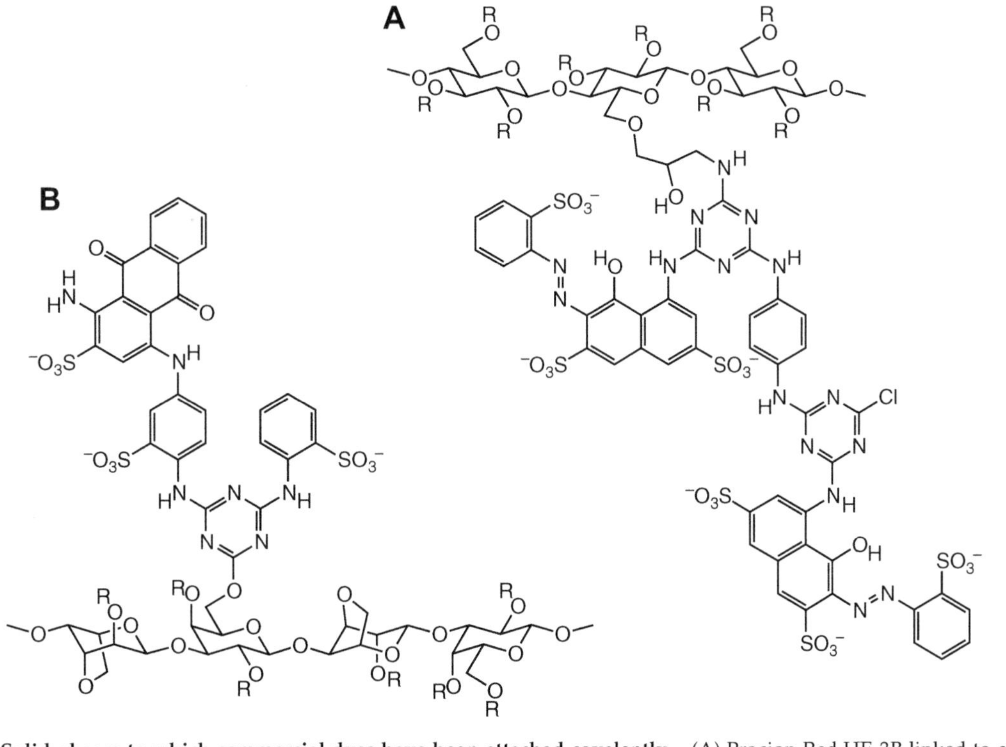

FIGURE 1.14 **Solid phases to which commercial dyes have been attached covalently.** (A) Procion Red HE-3B linked to cellulose, a polymer of glucose. (B) Cibacron Blue 3GA linked to cross-linked agarose, a polymer of the disaccharide 3,6-anhydro-α-L-galacto-β-D-galactan. The segment of agarose shown in the drawing does not have any of the cross-links, which are added to make the solid phase more rigid. Cross-linked agarose is also known as sepharose. In both drawings, the symbol R denotes positions in the polysaccharide occupied either by a hydrogen or by another molecule of the dye. Either of these solid phases can be used for either adsorption and elution or chromatography.

Experimental Plan

It is impractical to load the large volume of the pool of FNR from the anion exchange directly onto the solid phase to which the dye is covalently attached. Therefore, the sample is first concentrated by ultrafiltration under centrifugal force. Another advantage of this procedure for concentrating the enzyme is that it can be used to lower the ionic strength of the solution of protein.

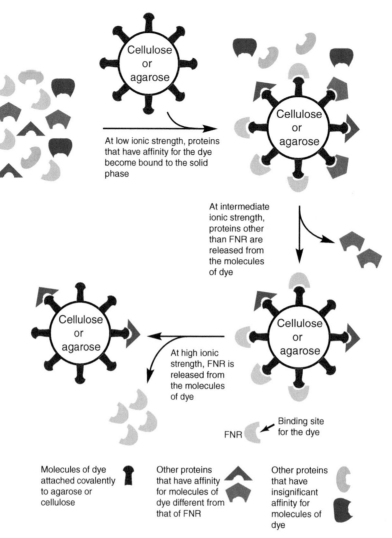

FIGURE 1.15 **Adsorption and elution of FNR from a solid phase to which a dye has been covalently attached.**   A mixture of proteins from chloroplasts, previously selected by anion exchange, is applied to the solid phase in a solution of low ionic strength. Only a few proteins, among them FNR, bind to the covalently attached molecules of dye. Proteins other than FNR are removed by washing with an excess of the solution of low ionic strength. The FNR elutes from the matrix as higher ionic strengths are applied.

Ferredoxin-NADP$^+$ reductase adsorbs to the dye on the solid phase when it is dissolved in a solution of low ionic strength (< 0.1 M) and can be eluted off the solid phase by increasing the concentration of KCl to increase the ionic strength. The pool from the anion exchange has an ionic strength greater than 0.1 M, so this must be lowered to less than 0.1 M before loading the sample onto the solid phase to which the dye is attached, so that the protein will adsorb to it.

The pool from the anion exchange is divided into several portions and loaded into special tubes for ultrafiltration that use centrifugal force to push the solution through a semipermeable membrane (Figure 1.16). During the centrifugation about 90% of the solution, and all of the molecules and ions with diameters less than 3 nm (the *filtrate*), pass through the membrane, which acts as a sieve. All macromolecules, in this case proteins, that have diameters greater than 3 nm (the *retentate*) are retained above the membrane in about 10% of the original volume. The retentate is recovered, diluted twofold with water, and then loaded into fresh devices for a second spin. After the second spin, the volume of the retentate should be about 2% of the original volume of the pool. The concentration of protein is increased in both spins. Although the concentration of

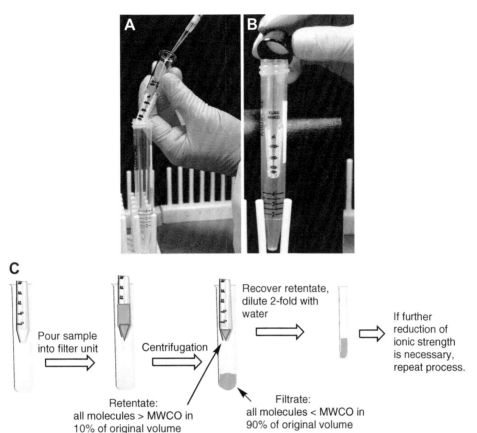

FIGURE 1.16 **Using a device for ultrafiltration under centrifugal force to increase the concentration of protein and lower the ionic strength.** (A) The sample is pipetted into the unit for ultrafiltration. (B) The unit is then inserted into the centrifuge tube. The Ultra-4 from Amicon is shown (4 mL maximum sample volume). A larger version of this device is also available. The molecular weight cutoff (MWCO) shown on the tube is an estimate of the largest spherical molecule of protein that will pass through the membrane. The tubes you will be using have micropores of an effective diameter of 3 nm, which is the diameter of a sphere of protein with a molar mass of $10,000 \ g \ mol^{-1}$, which, for historical reasons, are the units used by the manufacturer. (C) A schematic summary of the procedure.

NaCl does not change during either spin, the twofold dilution after the first spin results in a 50% decrease in the concentration of NaCl. If the pool initially had an ionic strength between 0.1 and 0.2 M, the 50% decrease will be sufficient. If the pool initially had an ionic strength between 0.2 and 0.3 M, another twofold dilution with water will be necessary to decrease the ionic strength to 25% of the initial value. The protein mixture now has an appropriate volume and ionic strength for adsorption to the solid phase.

Reagents, Supplies, and Equipment

Solid phase for adsorption:
  Cibacron Blue 3GA-agarose, type 3000-CL, in 0.5 M NaCl, 0.02% thimerosal Sigma-Aldrich. Or:
  Procion Red HE-3B-cellulose in 25% ethanol from Amocol Bioprocedures, Teltow, Germany. A packed volume of 3 mL is needed.

Solutions for adsorption and elution:
  Four stocks of 50 mM tricine, pH 8, with the following concentrations of KCl:
    50 mM KCl (75 mL).

0.1 M KCl (25 mL).

0.2 M KCl (25 mL).

0.5 M KCl (25 mL).

Reagents for the assay of the enzymatic activity of FNR (Zanetti, 1981):
The four solutions that were used for assaying the pools from DEAE.

Reagents for the Bradford assay (Boyer, 2012; Bradford, 1976):
The solution of Coomassie Brilliant Blue G-250 and the solution of the γ-globulin standard solution that were used for assaying the fractions from anion exchange.

Materials for adsorption of protein to the solid phase:
Glass chromatographic column with stopcock, a diameter of 1 cm, a height of 10 cm (Figure 1.17), and mounted on rack in cold room.

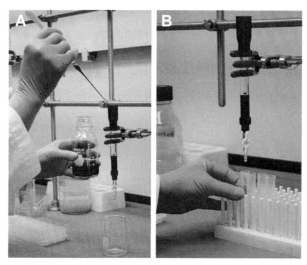

FIGURE 1.17   **Using a column filled with the solid phase containing covalently attached dye to purify FNR.**   (A) With the valve closed, a few milliliters of 50 mM KCl, 50 mM tricine, pH 8, is poured into the plastic column. A slurry of the solid phase is then pipetted in and more buffer added above the solid phase. The valve on the column is then opened to hasten the packing of solid phase at the bottom of the column. The column should never be allowed to run dry because this would result in air gaps that disrupt the flow of the solution through the bed. (B) To wash through undesired proteins and then elute FNR, buffers are added to the top of the column with a pipette. Because only a small number of fractions need to be collected, an automated fraction collector is not necessary; instead, fractions of the eluate are collected manually.

Supplies and equipment for the assays of enzymatic activity and concentration of protein:
Multiwell plate (Figure 1.13), cuvettes, or glass tubes similar to those used for assaying the DEAE fractions.

The same plate reader, conventional spectrophotometer, or Spec 20 that was used for assaying the DEAE fractions.

Supplies and equipment for ultrafiltration:
Amicon Ultra-15 10K Ultra Centrifugal Filter devices. These devices have a volume of 15 mL and a 10,000 g mol⁻¹ molecular weight cutoff (MWCO) (Figure 1.16 shows a smaller

version). You will need eight of these. Amicon also refers to 10,000 g mol^{-1} as the "nominal molecular weight limit." You should try to forget the term "molecular weight."

The same centrifuge and rotor that were used for the isolation and washes of the chloroplasts.

Precaution

Do not use the centrifuge without supervision.

The Bradford reagent contains phosphoric acid that can erode flesh and Coomassie Brilliant Blue G-250 that is toxic. Wear gloves.

Procedure (Two Laboratory Sessions)

PART 1. ULTRAFILTRATION UNDER CENTRIFUGAL FORCE FOLLOWED BY ADSORPTION TO THE SOLID PHASE

Use ultrafiltration to increase the concentration of protein and decrease the ionic strength (1.5 h).

Start with the pool from DEAE from the previous exercise.

Load up to 15 mL of the pool into each Amicon device (Figure 1.16 shows a version of the device with a smaller volume). You'll probably need about six of these devices. If you end up with an odd number, pair up with another group with an odd number of devices.

Spin at 5500 rpm (5000 × g) for 30 min at 4 °C in JA-14 rotor.

Pool the retentates from each device and record the volume, which should be about 10% of original volume.

Dilute twofold with water to lower the concentration of NaCl, then load into fresh Amicon devices. You'll probably need one or two of these devices.

Spin as before, then pool the retentates from each device and record the volume.

If the initial ionic strength of the pool was between 0.1 and 0.2 M, the sample is ready for adsorption onto the solid phase containing the dye. If the initial ionic strength was between 0.2 and 0.3 M, dilute the sample twofold with water, but do not concentrate further.

This is the *preadsorption sample.*

Remove two aliquots, 1 and 0.1% of the total volume.

It is imperative that you label each of the tubes with your initials, the date, and the type of tissue. There are two types of marking pens generally used in labs. The Sharpies are fine for assay tubes that are only needed for a single experiment, then discarded. For any tube that will be frozen in a slurry of dry ice in isopropanol you must use the alcohol-resistant Fisher or VWR laboratory markers! The isopropanol will dissolve the ink from other types of marker. Black ink sticks better than blue or other colors. Write directly on the tube; tape will fall off in the freezer.

• Freeze these samples rapidly in isopropanol and dry ice and store them at −20 °C. The 1% sample will be used for electrophoresis in FNR Exercise 4 and the 0.1% sample for the immunoblot in FNR Exercise 7.

• Store the remainder of the preadsorption sample in the cold room until you are ready to apply it to the column of the solid phase containing the dye.

Select FNR from the concentrated pool by adsorption to the solid phase (1.5 h).

• Prepare the solid phase (3-mL bed volume) in 50 mM tricine, 50 mM KCl (Figure 1.17).

• Before you add the preadsorption sample, wash the column with 10 mL of 50 mM tricine, 50 mM KCl, to remove all trace of the solution in which the solid phase has been stored on the shelf for long periods of time. This step is critical! The ethanol or the high salt in which the solid phase has been stored will interfere with binding of FNR.

• Load the preadsorption sample.

• Elute in steps using 10 mL of each of the three solutions of increasing concentration of KCl.

• Collect four fractions of 2 mL each at each concentration of KCl.

• After collecting the four fractions for one concentration of KCl, remove the excess of the solution at the top of the column and add the next solution.

• Keep these fractions in the cold room or on ice until you are ready to do assays for enzymatic activity.

PART 2. ASSAY FOR ENZYMATIC ACTIVITY AND FOR THE CONCENTRATION OF TOTAL PROTEIN IN THE ELUATE

Assay for activity of FNR in the fractions of the eluate from adsorption (2 h).

• Assay for enzymatic activity an aliquot from each of the fractions that you collected, as you did for the fractions from anion exchange. For accurate quantification, it is best that the initial dA_{490}/dt for a fraction be between 0.05 and 0.2 min⁻¹. If the rates of the assays for fractions with enzymatic activity are not in this range, repeat the assays with an appropriate change in volume. Again, it will be the case that several of the fractions will have little or no activity and will not need to be assayed accurately.

• Suggested volumes of each fraction to be assayed:

If using a plate reader: 40 μL of eluate in a 200 μL reaction.

If using a conventional spectrophotometer: 100 μL of eluate in a 0.5 mL reaction.

If using a Spec 20: 600 μL of eluate in a 3 mL reaction.

• Calculate the total activity (μmol min⁻¹) in each of the fractions.

- Most of the FNR activity should be present in the fractions of one of the four eluates with different ionic strengths, but significant activity may be detected at more than one ionic strength. Prepare a pool of the fractions for a given ionic strength that contain the majority of the FNR activity eluted at that ionic strength. If two or more different ionic strengths eluted significant portions of enzymatic activity, make a pool for each of them. Using the sum of the fractional activities and the final volume of the pools, calculate the concentration of enzymatic activity (μmol min^{-1} mL^{-1}) in the pools.

Quantify the concentration of total protein in the pools of FNR purified by adsorption (1 h).

Even though the amount of enzymatic activity that you retrieve in this experiment will be about the same as you retrieved from the DEAE solid phase (because the enzyme just binds to the solid phase and is then released), the amount of protein you retrieve will be much less than before because the enzyme has been purified. Therefore, you will have to use larger volumes in the Bradford assay for protein to get absorbances in the correct range. Assay aliquots of the pools with the Bradford assay as was done for the pools of the fractions from the anion exchange.

- First, prepare a standard curve with γ-globulin as before.

- Then assay the pools. To begin with, try the following volumes:

If you are using a plate reader, add 10, 20, or 50 μL of protein to 200 μL assays.

If you are using a conventional spectrophotometer, add 25, 50, or 125 μL of protein to 0.5 mL assays.

If you are using a Spec 20, add 150, 300, or 750 μL of protein to 3 mL assays.

- If necessary, repeat with different volumes to get absorbances within the range of the standard curve.

- Calculate the concentration of protein (mg mL^{-1}) and the specific activity (μmol min^{-1} mg^{-1}) of the pools.

- If you prepared more than one pool, use the measurements you've just made to decide, as you did before, which of these pools to combine and save for further experiments. Again, this is a judgment that you must make. Calculate the concentration of protein in the final pool and its specific activity.

This pool is the *FNR purified by adsorption.*

- If more procedures are scheduled to be performed in this laboratory session, keep the protein on ice. Otherwise, store in the cold room.

Analysis of the Data and Questions to Address in Your Report

1. *Sample added to the solid phase for adsorption (preadsorption sample).*
The measurements you made of the enzymatic activity of FNR and the concentration of protein from the previous exercise may be used to estimate the concentration of protein

(mg mL^{-1}), the total enzymatic activity (μmol min^{-1}), and the specific activity (μmol min^{-1} mg^{-1}) in the sample that you added to the solid phase for adsorption. Taking into account the increase in the concentrations that you performed before adding the sample to the solid phase, estimate these numbers for the sample that you added. This is only an estimate because you do not know how much protein and FNR you lost during the concentration.

2. *Pool of FNR purified by adsorption and elution (FNR purified by adsorption).*
   As for the fractions from anion exchange, you should note any color observed. Suggest an explanation for any color seen. Search the web for the extinction coefficient and the wavelength of maximum absorption for FNR. What color should the enzyme have? This color will be helpful in tracking FNR during the exercise in chromatography by molecular exclusion.

3. *Catalytic constant of the FNR that you have purified by adsorption and elution.*
   The catalytic constant (historically, this same rate constant has also been called the turnover number) is the number of molecules or number of moles of reactant converted to product by each active site or by each mole of active sites, respectively, every second (Berg et al., 2012; Voet and Voet, 2010). Because moles of reactant cancels moles of active sites, the unit for the catalytic constant is second^{-1}. Almost all enzymes are oligomers of two or more subunits, each of which contains an active site, but in some cases the oligomer includes both catalytic subunits, each of which contains an active site, and regulatory subunits, each of which binds metabolites that regulate the enzyme. The molar mass of each subunit of FNR, each of which contains one active site, is used to calculate the catalytic constant. The catalytic constant should be derived from the specific activity of the pure enzyme, but not from the specific activity of the protein from earlier stages of the procedure for purifying the enzyme. It is a fundamental property of an enzyme. If you know what the catalytic constant for the pure enzyme is supposed to be in the assay you are using, the catalytic constant you calculate for the current preparation gives you a sense of how pure it is at this stage.

   a. Obtain the molar mass of protein in a subunit of FNR from *S. oleracea* from the UniProt database (http://www.uniprot.org). This molar mass is calculated from the sequence of the amino acids in the mature protein (after the transit peptide has been removed). You should recall that the protein has a prosthetic FAD, so the actual molar mass of each subunit of FNR is greater than the molar mass of its protein, but you have only measured the mass of protein in the assays.

   b. Use the molar mass of protein in a subunit of FNR (g mol^{-1}) and the specific activity (μmol min^{-1} mg^{-1}) of the purified enzyme from the adsorption to calculate the catalytic constant (s^{-1}).

4. Discuss whether the measured catalytic constant convinces you that the enzyme being assayed is in fact FNR.

5. *Table summarizing the purification.*
   This table shows the progress of the purification at each step. The format should be similar to the template below (Table 1.1). In each case, you should use the numbers for the final pools from each stage of the purification. For calculation of the total volume

TABLE 1.1   Summary of the Purification of FNR from S. *oleracea*

	Total Volume (mL)	Total Protein (mg)	Concentration of Protein (mg mL^{-1})	Total Activity (μmol min^{-1})	Concentration of Activity (μmol min^{-1} mL^{-1})	Specific Activity (μmol min^{-1} mg^{-1})	Fold Purification	Recovery (%)
Anion exchange								
Adsorption and elution								

after each step of purification you must correct for removal of aliquots for the various analytical procedures. Some aliquots were removed for assays of enzymatic activity or the concentration of protein (performed immediately), and others were removed and stored for use in electrophoresis experiments in later exercises. It is critical to make this correction in the total volume after both steps of the purification scheme because the total volume is used for calculation of total protein, total activity, and percent recovery. Removal of aliquots at an earlier step of purification must be considered, in addition to those from the most recent step. For example, if 2% of the pool from anion exchange was removed for various assays and 4% of the pool from adsorption and elution was removed, the total volume of the pool from anion exchange should be corrected by dividing by 0.98. The volume of the pool from adsorption and elution should be corrected by dividing by 0.98 and then by 0.96.

6. An increase in the specific activity of an enzyme after each step of a purification verifies that these methods are effective. Is the specific activity of FNR after adsorption and elution significantly higher than after the anion exchange? Does it appear that adsorption and elution is an effective method for purification of FNR?

7. Does the measured specific activity after the adsorption and elution convince you that the FNR in the final pool is pure? If not, what experiment is now needed to verify purity?

8. Consider the differences between anion exchange and adsorption and elution; for example, the differences in the availability of an appropriate solid phase, the maximum amount of a cellular lysate or sample of protein that may be applied to a given amount of solid phase, the method for dissociating the bound enzyme from the solid phase, and any other factors. Which type of solid phase is more specialized? Why was the anion exchange used prior to adsorption? What would be the disadvantage of applying the lysate from the chloroplasts directly to the solid phase for adsorption?

FNR EXERCISE 4

Electrophoresis of Proteins from the Chloroplasts and Purified FNR on Gels of Polyacrylamide Cast in Solutions of Dodecyl Sulfate

Technical Background

Assays of enzymatic activity at the various stages of purification do not provide a direct measurement of purity. Electrophoretic separation, followed by the visualization of all of the proteins in the sample with a nonspecific stain, is the accepted technique for determining what proteins other than the protein of interest are present and hence if the enzyme is pure. If the proteins separated by the electrophoresis are to be assayed for their enzymatic activity after the electrophoretic separation, nondenaturing conditions are used. Electrophoresis of proteins unfolded by dodecyl sulfate, however, provides more information about the proteins that are present as well as the length of their constituent polypeptides.

The polypeptides that have folded to form the proteins in the sample are all unfolded by causing them to bind the detergent dodecyl sulfate, usually used as its sodium salt, sodium dodecyl sulfate. The complexes between the molecules of protein and dodecyl sulfate are then separated by electrophoresis on a gel of polyacrylamide. The acronym PAGE is often used to refer to the method of polyacrylamide gel electrophoresis (Boyer, 2012; Davis, 1964; Kyte, 2007, pp 421–423; Laemmli, 1970; Ornstein, 1964; Shapiro et al., 1967; Tulchin et al., 1976; Weber and Osborn, 1969).

To form saturated complexes between ions of dodecyl sulfate and the constituent polypeptides of the molecules of protein, sodium dodecyl sulfate is added to the solution of protein to a ratio greater than 2 μg for every microgram of protein, and to a final free concentration of sodium dodecyl sulfate of at least 0.1%. 2-Sulfanylethanol, which reduces any disulfides in the protein, is added to a final concentration of 1%, and the resulting sample is placed in a bath of boiling water. The sample is brought to 100 °C to ensure that the proteins in the sample, in particular any endopeptidases that might be present, are unfolded rapidly. Endopeptidases prefer to digest unfolded proteins, and they are usually more resistant to unfolding by dodecyl sulfate than other proteins, so if the unfolding is allowed to proceed slowly, there is a possibility that some of the unfolded polypeptides will be cleaved by endopeptidases before the endopeptidases themselves succumb by unfolding.

Molecules of protein that are soluble in aqueous solution, such as FNR, bind 0.54 ± 0.01 molecules of dodecyl sulfate for every amino acid in their sequence. The constancy of this ratio is the reason that the procedure can be used to estimate the length of a particular polypeptide. The structure of the complex that forms between ions of dodecyl sulfate and a polypeptide is not known in detail, but what is known is that the polypeptide becomes completely unfolded from its native structure (Figure 1.18). Sodium dodecyl sulfate is also included in the gel and the running buffer, to ensure that the polypeptides remain saturated with dodecyl sulfate during electrophoresis. Because subunits in an oligomer are held together by specific interactions that depend on the maintenance of the tertiary structure of the subunits, when the subunits of an oligomer unfold, they also separate from each other and become independent polypeptides, coated with dodecyl sulfate. Thus, enzymes that function as oligomers in their natural state are turned into their individual polypeptides no longer associated with each other in a solution with sufficient dodecyl sulfate to saturate them. Because polypeptides to which ions of dodecyl sulfate are bound are anionic, they migrate toward the anode (positive terminal) of the apparatus for electrophoresis.

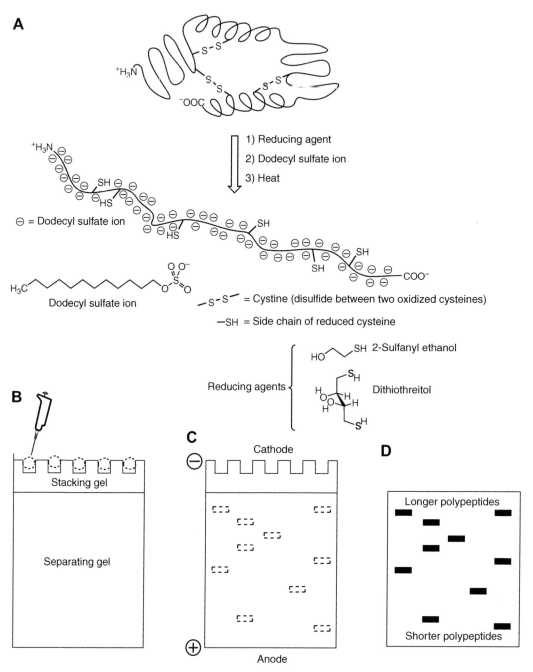

FIGURE 1.18 **Unfolding of polypeptides by dodecyl sulfate and separation of the complexes of polypeptides and ions of dodecyl sulfate by electrophoresis on gels of polyacrylamide.** (A) All of the secondary and tertiary structure of a polypeptide is disrupted during its saturation with dodecyl sulfate. The ions of dodecyl sulfate bind to the polypeptide, and as a result, at saturation, the polypeptide is coated uniformly with anionic dodecyl sulfates. Any disulfides present in the protein are cleaved by reduction with either 2-sulfanylethanol or dithiothreitol. (B) The *stacking gel* is buffered at pH 6.8 and typically contains 4.5% by mass of polyacrylamide, while the *separating gel* is buffered at pH 8.8 and can contain anywhere between 8 and 15% by mass of polyacrylamide, depending on the lengths of the polypeptides that are being separated. The two parts of the gel both contain 0.1% sodium dodecyl sulfate but differ in the identity of the ions, the concentration of the ions, and the pH of the solution within the gels. Their concentrations of polyacrylamide differ because the stacking gel should not impede the movement of the polypeptides through it. Consequently, it should be of the lowest concentration of polyacrylamide that can form a gel. Samples are loaded into the wells formed within the stacking gel. (C) A voltage is applied between cathode and anode. In response to this voltage, an electrical current passes through the ionic solution within the gel between two electrodes: one submerged in the solution in the reservoir at the cathode and one submerged in the solution in the reservoir at the anode. It is the voltage that has been applied, however, not the current, that separates the complexes between the polypeptides and ions of dodecyl sulfate. The polypeptides, coated with ions of dodecyl sulfate, migrate towards the anode. They initially enter the stacking gel, in which a sharp, stable moving boundary between two different solutions is created. All of the complexes between the polypeptides and ions of dodecyl sulfate are trapped in this sharp, moving boundary and become compressed into a tight band. When they reach the top of the separating gel, the mobility of the moving boundary abruptly increases, and it leaves the polypeptides at the top of the separating

The free electrophoretic mobilities of the polypeptides of all soluble proteins when they are saturated with dodecyl sulfate, their electrophoretic mobilities in the absence of a gel of polyacrylamide, are all identical to each other. Consequently, it is only the sieving (Kyte, 2007, pp 421–423) accomplished by the polymeric network of the polyacrylamide that separates one polypeptide from another. The rate of migration of each individual polypeptide saturated with dodecyl sulfate through the polymeric network of the gel of polyacrylamide is a function of only the surface area of the complex. Because it has also been shown experimentally that this rate of migration is a function only of the length of the polypeptide, whatever structure the complex between a polypeptide and ions of dodecyl sulfate assumes, it must be one that has a constant diameter and increases proportionally in length as the polypeptide becomes longer.

The relative mobility of a particular complex between a polypeptide and ions of dodecyl sulfate is the ratio between the rate at which that complex migrates through the gel of polyacrylamide and the rate at which a standard migrates through the gel of polyacrylamide. Bromophenol Blue is the standard that is almost always used on a polyacrylamide gel cast in a solution of dodecyl sulfate. It is a dye that is probably enclosed within unconnected individual micelles of dodecyl sulfate and has the migration of those micelles. These micelles are smaller than and move through the gel more rapidly than any of the complexes between polypeptides and ions of dodecyl sulfate unless the polypeptide is so short that it is also completely enclosed in a micelle of dodecyl sulfate, in which case it has the same mobility as the Bromophenol Blue. The symbol R_f, that stands for *relative to the front*, is often used to represent the relative mobility. Historically, the front referred to is the front of the solvent moving through the chromatographic paper in the method of paper chromatography. It is in the application of this technique that the term relative mobility arose. The symbol R_f was then applied to any method of separating molecules on a field. The front, in the case of electrophoresis in discontinuous buffer systems, is the rapidly descending stable boundary between two ionic solutions in which micelles of dodecyl sulfate and Bromophenol Blue are trapped. This stable boundary moves in front of all of the complexes between polypeptides and dodecyl sulfate.

A plot of the natural logarithms of the relative mobilities of the complexes of a standard set of polypeptides saturated with ions of dodecyl sulfate is a linear function of the lengths of the polypeptides. Because the running length of a polypeptide is directly proportional to the number of amino acids it contains, the length of a polypeptide is usually expressed in terms of the length of its sequence of amino acids. A mixture of complexes between a set of standard polypeptides and ions of dodecyl sulfate is run on one gel and the natural logarithms of their relative mobilities are plotted as a function of the lengths of their sequences of amino acids to obtain a standard line. The relative mobility of a polypeptide of unknown length is used with this standard curve to estimate its length.

The Coomassie Brilliant Blue G-250 (Figure 1.10) that was used in the Bradford assay is also used to stain the gel to locate the positions of the various polypeptides. Because all of the polypeptides from the same homooligomer travel with the same electrophoretic mobility, the stained gel contains a band of color for each type of polypeptide. The minimum mass of protein needed to

---

gel, no longer trapped in the moving boundary. Each of the polypeptides is now free to move through the network of the polyacrylamide in the separating gel at its own characteristic pace. This is the process of sieving that separates the complexes based on the length of their polypeptides. (D) After the separation is completed, the polypeptides are visualized in the separating gel by staining them with Coomassie Brilliant Blue.

produce a visible band of stain is around 2 μg. If more than 10 μg of one protein is loaded, the band of staining becomes wider, and it is difficult to get an accurate measurement of its migration.

Experimental Plan

As was done to the pool from DEAE, the concentration of protein in the FNR purified by adsorption is increased by ultrafiltration under centrifugal force. This makes it easier to load an appropriate mass of protein onto the gel of polyacrylamide. On the electrophoresis, you will compare the following samples: (1) a mixture of standard polypeptides of known length, (2) the preadsorption sample of the proteins in the material loaded onto the solid phase for adsorption, and (3) a sample of the FNR purified by adsorption in which the concentration of protein has been increased by ultrafiltration under centrifugal force.

Reagents, Supplies, and Equipment

Reagents for preparation of protein samples:
   10% 2-sulfanylethanol, 50 mM sodium phosphate, pH 7.2.

   80% glycerol, 0.25% Bromophenol Blue.

   10% sodium dodecyl sulfate (100 μg μL^{-1}).

   Mixture of standard polypeptides: Sigma-Aldrich item M-3913. The polypeptides are already coated with dodecyl sulfate. Reconstitute in water at 2 μg μL^{-1}.

Reagents for preparation of the gel of polyacrylamide, for electrophoresis, and for staining:
   0.4% sodium dodecyl sulfate, 0.5 M Tris chloride, pH 6.8, for the stacking gel.

   0.4% sodium dodecyl sulfate, 1.5 M Tris chloride, pH 8.8, for the separating gel.

   1% sodium dodecyl sulfate, 1.92 M glycine, 0.25 M Tris, pH 8.8, for the electrophoresis reservoirs. This stock solution will be diluted tenfold to make the solutions added to the reservoirs.

   30% acrylamide, 0.8% bisacrylamide.

   10% (m/v) ammonium persulfate to catalyze the polymerization of the acrylamide. Prepared on the day it will be used.

   N,N,N′,N′-tetramethyl-1,2-ethanediamine to catalyze the polymerization of the acrylamide. A liquid reagent also known as tetramethylethylenediamine or TEMED.

   150 mL Invitrogen Simply Blue stain: Coomassie Brilliant Blue G-250. Although it is not in an organic solvent, the stain itself is toxic, so wear gloves.

   15 mL 20% NaCl, which is needed if overnight staining will be done.

Supplies for ultrafiltration, preparing the gel of polyacrylamide, and for staining:
   Amicon Ultra-4 10K ultrafiltration devices.

   The materials needed for preparation of the gel are shown in Figure 1.19.

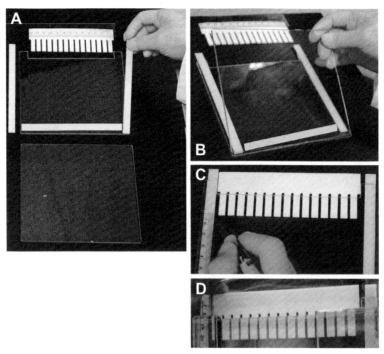

FIGURE 1.19 **Casting a gel of polyacrylamide.** (A) The materials. (B) The spacers are placed along the sides and bottom of the notched plate. The unnotched plate is then set down on the spacers, its edges aligned with the notched plate. The edge of the assembly containing the wide notch will be at the top of the gel. (C) The comb is inserted at the top, between the plates. Approximately 90% of each tooth should be below the edge of the wide notch in the front plate. The height of the stacking gel, which begins at the bottom of the teeth, should be about 10% the height of the separating gel (see Figure 1.18), so 90% of the height of the gel from the bottom of the teeth to the bottom of the mold should be reserved for the separating gel, and 10% for the stacking gel. A mark is made on the notched plate at this location to indicate the desired upper edge of the separating gel. The comb, which was only inserted for this measurement, is then removed and the assembly of the plates and the spacers is secured with metal clamps and set upright. *Critical*: **The periphery of the assembly is sealed with molten agar or electrical tape to prevent leakage**. The solution for the separating gel is poured between the plates, up to the mark on the notched plate. The solution is overlaid with ethanol. After polymerization occurs (allow about 20 minutes) the ethanol is poured off. The stacking gel can be formed now or the separating gel can be stored for later use. If the stacking gel is to be poured now, the solution for the stacking is poured in on top of the separating gel. (D) Immediately after pouring in the solution for the stacking gel on top of the separating gel, the comb is inserted to the same depth as before. The assembly is left undisturbed until the stacking gel has polymerized. An outline of the wells should be made on the unnotched plate with a marking pen to aid in loading the sample, because you may have a hard time seeing the wells made by the teeth once the buffer for the upper reservoir is poured on.

Wash bottle of 95% ethanol.

Plastic tray with lid for staining and destaining.

Equipment for ultrafiltration, electrophoresis, and staining:
   Centrifuge and rotor as used for the isolation and washing of the chloroplasts.

Apparatus and power supply for electrophoresis (Figure 1.20).

Microwave oven.

Precaution

Do not use the centrifuge without supervision.

Acrylamide is a neurotoxin that can be adsorbed through the skin: wear gloves and use pipette bulbs. Never pour acrylamide down the drain. Polymerized acrylamide is much less toxic. Polymerize any unused acrylamide and then toss the solid into the trash.

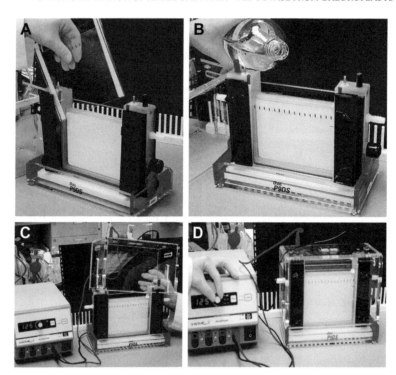

FIGURE 1.20 **Assembly of the apparatus for vertical electrophoresis.** (A) The assembly of the glass plates that formed the mold for the gel is loaded into the apparatus for the electrophoresis. (B) Running buffer is poured into both the inner chamber, where it provides an electrical conductor between the wire of the cathode and the top of the gel, and into the bottom reservoir, where it provides an electrical conductor between the wire of the anode and the bottom of the gel. (C) Electrical cords are permanently attached to the top of the cover for the electrophoretic chamber. When the gel is ready, the red (+, anode) and black (−, cathode) cords are connected to the power supply. **Before loading the samples**, you should lower the top of the chamber onto the apparatus to make sure that an electrical current registers on the meter when the voltage is applied, to check the continuity of the electrical circuit. (D) After an appropriate current is detected, the power supply is put in standby mode, which turns off the voltage, the cover is removed, and samples are prepared for loading.

Procedure *(One Laboratory Session)*

Increase the concentration of the FNR purified by adsorption by ultrafiltration under centrifugal force (30 min).

- Start with the pool of FNR purified by adsorption.

- Decrease the volume of the pool to ~10% of its original value as you did before. Each Amicon device holds up to 4 mL (Figure 1.16). You may need to use two.

- Spin at 7000 rpm (7500 × *g*) for 20 min at 4 °C in JA-14 rotor (or equivalent).

- Recover the retentate from the device or the two devices, transfer to a test tube of the appropriate size and record its final volume. This is the *concentrated FNR purified by adsorption.*

- Use the data from the assays for enzymatic activity of FNR and concentration of protein in the unconcentrated pool to estimate the enzymatic activity and concentration of protein in the sample after the concentration.

- Divide the concentrated solution of enzyme into several different portions:

1. For the electrophoresis of the polypeptides coated with dodecyl sulfate: The procedure requires 32 μg of protein, but you should remove an aliquot of 64 μg in case something goes wrong and you have to repeat the experiment. Keep this portion on ice until you are ready to prepare the samples for loading the gels of polyacrylamide.

2. Kinetic assay to determine the Michaelis constant, K_{mNADP}, of FNR for NADP (if this exercise is on the course schedule): Remove a portion of the pool that contains 0.4 μmol min^{-1} of enzymatic activity.

3. Immunoblot (if this is on the course schedule): Remove a portion containing 1 μg of protein.

4. Chromatography by molecular exclusion: The remainder of the pool (there should be about 1.6 μmol min^{-1} of enzymatic activity) will be used for this exercise.

Label those portions that will not be used immediately, and freeze them rapidly in a mixture of isopropanol and dry ice. Store them at −20 °C.

Prepare the separating gel (40 min).

Have the mold for the cast (Figure 1.19) ready before adding the catalysts, ammonium persulfate and tetramethylethylenediamine, to the following mixture:

30% acrylamide, 0.8% bisacrylamide	12.5 mL
0.4% sodium dodecyl sulfate, 1.5 M Tris chloride, pH 8.8	7.5 mL
Water	9.6 mL
10% (m/v) ammonium persulfate	0.40 mL
Tetramethylethylenediamine	0.015 mL
Total	**30 mL**

Add the ingredients in the order of the list. The final solution should be poured between the plates **immediately** after the catalysts have been added. The gel solution is then overlaid with 95% ethanol to prevent interference from oxygen and to ensure a flat top. The concentration of polyacrylamide in the final separating gel is nominally 12.5%. Allow the gel to sit for 20 min to permit full polymerization. While it is polymerizing you should prepare the samples of protein for the electrophoresis (see next section). This (lower) part of the gel may be prepared before the day of use. For storage after polymerization: cover with buffer for the separating gel (0.1% sodium dodecyl sulfate, 0.375 M Tris chloride, pH 8.8) and place the separating gel in the cold room to prevent it from drying out at the top.

Prepare samples for electrophoresis (20 min).

The guide in Table 1.2 specifies the amounts of each sample to prepare for the electrophoresis. In addition to the sodium dodecyl sulfate, the 2-sulfanylethanol, and the Bromophenol Blue, glycerol is added to increase the density of the sample and facilitate loading.

The reagents to be added to these samples of protein are summarized in Table 1.3. Add the water, the 2-sulfanylethanol, and the sodium dodecyl sulfate, in that order, to bring the sample to 40 μL. Put the sample in the bath of boiling water for 1 min as soon as the dodecyl

TABLE 1.2   Preparation of Samples for the Electrophoresis of the Proteins from the Chloroplasts[a]

	Concentration of Protein ($\mu g\ \mu L^{-1}$)	Volume (μL)	Mass of Protein (μg)
Standard polypeptides[b]	2	8	16
Preadsorption sample			50
			100
Concentrated FNR purified by adsorption			2
			10
			20

[a] *The table should be copied into your notebook and should be completed after the concentration of protein has been estimated following the ultrafiltration.*
[b] *If the recommended standards from Sigma-Aldrich are used, they have already been saturated with dodecyl sulfate, so they are ready to load and the sample does not need to be brought to 100 °C.*

TABLE 1.3   Reagents to Be Added to Each Sample of Protein for Electrophoresis

	Volume (μL)	Final Concentration
Protein sample	Variable	Variable
Water	Variable	
10% 2-sulfanylethanol, 50 mM sodium phosphate, pH 7.2	5	1% 2-sulfanylethanol, 5 mM sodium phosphate
2% (20 $\mu g\ \mu L^{-1}$) sodium dodecyl sulfate[a]	Variable	0.1% free concentration
0.25% Bromophenol Blue, 80% glycerol	10	0.05% Bromophenol Blue, 16% glycerol
Total	50[b]	

[a] *Calculate to be sure that at least 2 μg of sodium dodecyl sulfate has been added for every microgram of protein in addition to the amount of sodium dodecyl sulfate needed to make the final concentration 0.1%.*
[b] *If the wells on the gel form properly, as much as 150 μL may be loaded. If it is necessary to have a sample with a volume greater than 50 μL, make appropriate increases in the amounts of the other components.*

sulfate has been added to rapidly unfold all of the proteins. After heating to 100 °C and cooling the sample, add the Bromophenol Blue and glycerol. Spin the final samples in a tabletop centrifuge briefly to pull all of the liquid to the bottom of the tube. The samples may be prepared before the day of the electrophoresis and stored in the cold room. Some of the sodium dodecyl sulfate may precipitate in the cold, but it will redissolve when the sample is brought to room temperature.

Prepare the stacking gel (25 min).

This upper part of the gel should be prepared in the same laboratory session in which electrophoresis is performed (Figure 1.19D). This ensures that the concentrations of the various ions and the pH of the solutions in the separating gel and the stacking gel do not have the opportunity to be altered by diffusion across the boundary of the two gels. If the boundary of ionic concentration and pH between the two parts of the completed gel are not sharp when samples are loaded, the mobility of the moving boundary will not increase dramatically at the top of the separating gel, and if the two solutions diffuse together almost completely, the moving boundary itself will not form to trap the complexes at its interface.

- The final concentration of polyacrylamide in the stacking gel is 4.5%, which is the minimum concentration required to create a gel. The gel is formed to prevent convection from occurring when the voltage is applied, which would rapidly mix the stacking solution and the solution layered on top of the sample, and defeat the purpose of a discontinuous electrophoresis. The minimum concentration of polyacrylamide is used to prevent any sieving and the resulting separation of the complexes between dodecyl sulfate and the polypeptides from occurring in the discontinuous electrophoresis.

30% acrylamide, 0.8% bisacrylamide	1.5 mL
0.4% sodium dodecyl sulfate, 0.5 M Tris chloride, pH 6.8	2.5 mL
Water	5.9 mL
10% (m/v) ammonium persulfate	0.10 mL
Tetramethylethylenediamine	0.020 mL
Total	**10 mL**

Again, do not add the catalysts ammonium persulfate and tetramethylethylenediamine until you are ready to pour the mixture into the mold and slide the comb into it.

Prepare the assembly for electrophoresis (10 min).

The next two figures and the following instructions are written for the *Owl Dual-gel Vertical Electrophoresis System* marketed by Thermo Scientific (http://www.thermoscientific.com). There are many variations of this type of system, and they all function in a similar manner.

Electrophoresis on polyacrylamide gels that are cast in solutions of dodecyl sulfate is performed at room temperature because sodium dodecyl sulfate precipitates from solution at 4 °C. Electrophoresis in the absence of dodecyl sulfate is usually performed in a cold room to decrease the heating caused by the electric current flowing through the resistance of the solution.

- Prepare 800 mL of running buffer containing 192 mM glycine, 25 mM Tris, pH 8.8, by diluting the stock solution tenfold.

- Mark the borders of the sample wells on the unnotched plate with a Sharpie pen (Figure 1.21) before removing the comb to make it easier to load the samples.

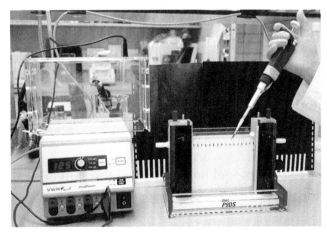

FIGURE 1.21 **Loading a vertical gel.** The markings on the unnotched (front) plate show the outline of the wells for the samples. The glycerol in the solution containing the sample (Table 1.3) increases its density so it will sink through the electrophoretic buffer. The tracking dye allows you to follow this underlaying. It is best to use a P20 pipette for smooth application (multiple additions are loaded for any well requiring more than 20 μL).

- Remove the comb that formed the wells in the stacking gel, and load the assembly of the glass plates into the apparatus for electrophoresis (Figure 1.20A). Secure the gel assembly so that buffer poured into the inner chamber will not leak out.

- Pour enough buffer into both reservoirs so that each of the electrodes is covered and full contact is made with both the top and the bottom of the gel (Figure 1.20B).

- Place the top on the apparatus so that electrical contacts are established, and turn on the voltage (Figure 1.20C). Set voltage at 125 V and check for current; expect 10–100 mA (Figure 1.20D). The display on the power supply will show a voltage even if no circuit is established. A reading of 0 mA indicates a lack of an electrical circuit and a reading of much greater than 100 mA indicates a short circuit. Try to spend as little time as possible (less than a minute) in this check for continuity, because the current will begin to change the ionic compositions of the boundaries and decrease the effectiveness of the stacking.

- As soon as possible, put the power supply on standby to turn off the voltage. Turning off the voltage also eliminates the risk of electrical shock when you open up the apparatus. Remove the top of the apparatus.

Load and perform electrophoresis (1.8 h).

- Load the samples into wells by submerging the tip of the pipette into which you have drawn them and slowly releasing each sample so it sinks into the bottom of the well (Figure 1.21). Place the top on the apparatus. Start the electrophoresis at 125 V, constant voltage, and continue until the tracking dye passes the top of the separating gel (around 30 min) and the complexes of the polypeptides and the ions of dodecyl sulfate have been dumped on the top of the separating gel.

- Increase the voltage to 250 V for the actual electrophoretic separation of the different polypeptides and continue until dye is within 1 cm of the bottom edge of the separating gel (about 60 min). Make a record of the total run at 250 V in units of volt-hours. The reason that a lower voltage is used for stacking than for separating is that, until the moving boundary enters the separating gel, the boundary itself creates a significant resistance. If the higher voltage were applied during stacking, the sample might overheat and significant convection would occur, which would destroying the stacking.

Remove the gel from the mold and prepare it for staining (10 min).

- Obtain a plastic tray and add 200 mL of water.

- Remove the spacers and pry apart the plates.

- Remove and discard the stacking gel because all of the polypeptides should be in the separating gel.

- Slide your thumbs under the corners of the separating gel and lift it off of the plate.

- Measure the length of the gel and the distance traveled by the tracking dye. This is necessary because the dimensions of the gel may change during the stain and destain procedure, and the tracking dye will diffuse out of the gel. These measurements will be used in the calculation of the relative mobilities of the polypeptides as explained below.

- Place gel in the water in the tray.

Stain and destain the gel.

For a 12.5% acrylamide gel that has separated complexes of polypeptides and dodecyl sulfate, either of the following two methods of staining may be used:

Quick method using a microwave oven (40 min):

- Wash in water three times for 5 min each time.

- With the gel in water, heat in the microwave oven until the solution is 75 °C. This takes about 4 min. Watch carefully to avoid boiling.

- Pour off water.

- Add fresh water and then repeat the heating in the microwave oven two more times, changing the water each time.

- Add 150 mL Simply Blue and heat in the microwave oven as before.

- Allow the solution to cool at least 5 min at room temperature, then begin destaining in water at room temperature.

Overnight method at room temperature. (20 min to prepare gel for overnight incubation):

For this method, NaCl is added to the Simply Blue to a final concentration of 2%. It is best to use a plastic tray that has a lid to prevent evaporation. Alternatively, Saran Wrap or aluminum foil may be used to cover the tray.

- Wash the gel three times for 5 min each time in water.

- While you are washing the gel, mix in a beaker or flask 135 mL of Simply Blue with 15 mL of 20% NaCl.

- Submerge the gel in the solution of stain, cover the tray, and leave the covered tray in the fume hood until your next scheduled session.

- Discard staining solution in the hazardous waste container, then rinse the gel twice for 5 min each time with water.

- Leave the gel in the water for several hours or overnight to clarify the gel surrounding the bands of stain. You may make an image of the gel as soon as all of the major bands of dye can be seen, but after you have made the photograph, return the gel to the water and leave it in the water long enough to remove all of the unbound stain from the gel. When the gel is cleared of dye and transparent where there are no bands, you will be able to see the bands for all of the polypeptides present in each sample, no matter how faint they are.

Make an image of the gel (10 min).

- Lay the destained gel on a glass plate and place it on the light box.

- Lay a clear plastic ruler alongside the lane containing the standards with the mark for 0 cm at the top of the separating gel, which is the origin of the separation.

- Adjust the zoom and focus on the lens of the camera.

- Optimize contrast by adjusting the exposure time or the aperture on camera lens or both of them.

- Capture the image in the default format for the camera.

- Export the image to JPG format so that it can be opened by the usual applications.

Analysis of the Data and Questions to Address in Your Report

1. Obtain the length in number of amino acids for each of the standard polypeptides from the UniProt database (http://www.uniprot.org).

2. Calculate the relative mobility, R_f, of each of the stained bands representing a different polypeptide. The relative mobility is the distance traveled by the polypeptide divided by the distance traveled by the tracking dye in that lane. If the gel has changed its dimensions during the staining, correct the position of the tracking dye that you measured on the unstained gel for the change in the length of the gel.

3. A semilog plot of the relative mobilities of the standard polypeptides should be created. Calculate the negative of natural logarithm for the relative mobility of each standard. Plot these values on the y axis as a function of the lengths of the respective polypeptides on the x axis.

Hint: The relative mobilities will all be between 0 and 1. Thus, you will be plotting the negatives of natural logarithms of numbers between 0 and 1. In this case, a semilog plot generated by spreadsheet software with a logarithmic axis may be difficult to interpret. So it is best to calculate the negatives of the natural logarithms with a calculator and then plot them on a linear scale. You can also use a spreadsheet to calculate the negatives of the natural logarithms and then have a graphing program plot the values.

4. Estimate the length of the major polypeptide in the lane of the sample of FNR purified by adsorption by interpolation on the semilog plot. If the plot has obvious curvature, draw a curve through the points and interpolate using the curve rather than a straight line. The stacking process can alter the mobilities of the standards, causing the natural logarithms of their mobilities to be a nonlinear function of their length. Is the estimated length of the major polypeptide close to the actual length of the constituent polypeptide of FNR from *S. oleracea* (see http://www.uniprot.org; remembering to use the length of the polypeptide from the mature protein)? Does this suggest that it may actually be the polypeptide of FNR?

5. Does the electrophoresis provide any clues as to the quaternary structure of FNR in the chloroplast?

6. What might cause the specific activity of the FNR purified by adsorption to be altered by the final step in which the protein is concentrated by ultrafiltration?

FNR EXERCISE 5

Kinetic Assay for the K_m of FNR for NADPH

Technical Background

When the kinetics of an enzyme are measured, the conditions are set up so that the equilibrium catalyzed by the enzyme is forced to proceed in only one direction. The direction chosen by the investigator defines what the reactants and products of this artificial unidirectional reaction will be. The reactants are the substrates for the enzyme that are decreasing in concentration and the products are the substrates for the enzyme that are increasing in concentration as the reaction progresses. A Michaelis constant, K_m, is a kinetic constant that in most instances refers to only one of the reactants in a kinetic experiment. In an experiment to measure the Michaelis constant K_{mA} for a particular reactant A, the concentration of that reactant is varied while the concentrations of all of the other reactants are held constant, usually at saturating concentrations.

In the kinetic assay of an enzyme that catalyzes oxidation and reduction using NAD or NADP, there are typically two reactants and two products because no matter which direction is chosen for the experiment, the enzyme catalyzes the transfer of a formal hydride ion from one molecule, the reductant, to another molecule, the oxidant. The only substrates, in each direction, are the respective oxidant and reductant. As in the assays of activity performed in the previous exercises, the direction chosen for the kinetic experiments you will be performing with FNR is the oxidation of NADPH, the reductant, and the reduction of INT, the oxidant, rather than, for example, the reduction of $NADP^+$ by ferredoxin.

The most informative Michaelis constant for a particular reactant is that observed for the reaction when the concentrations of the other reactants are all at saturation; in other words, at levels where changes in their concentrations affect neither the rates of the enzymatic reaction nor the observed Michaelis constant for the reactant being varied. The reported Michaelis constant, K_{mINT}, of FNR for INT is 0.125 mM (Zanetti, 1981). Consequently, for the kinetic measurements you will be performing, the concentration of INT will be 0.5 mM, close enough to saturation to satisfy this requirement.

Experimental Plan

The kinetic assay for the initial rates of the enzymatic reaction will be similar to that used when you were assaying samples for the activity of FNR. Aliquots of the concentrated FNR purified by adsorption will be assayed. The concentration of FNR in the concentrated FNR purified by adsorption should have been increased tenfold by the ultrafiltration under centrifugal force that you performed after the purification. An aliquot containing 0.4 μmol min^{-1} of enzymatic activity should have been reserved for this experiment. You will vary the molar concentration of NADPH from 0 to 0.2 mM. As in the earlier exercises, the formazan from the reduction of INT by the prosthetic $FADH_2$ on the enzyme will be quantified by the increase in A_{490}. The reaction for 0 mM NADPH is a negative control to determine if there is any contaminant in the preparation that is capable of reducing INT in the absence of NADPH, which FNR is unable to do. The dA_{490}/dt will be converted to an initial velocity, v, and Equation i.2 will be fit to the data to determine, as a parameter for the fit, the Michaelis constant, K_{mNADPH}, for NADPH.

Reagents, Supplies, and Equipment

Reagents for the kinetic assays:
　　5 mM NADPH, 50 mM sodium bicarbonate, pH 9. Minimize exposure to air.
　　5 mM INT, 70% ethanol, 5% Triton detergent.

5 M NaCl.
1 M Tris chloride, pH 9.

Appropriate tubes, cuvettes, or multiwell plate for spectrophotometer.

Assorted pipettes.

Plastic, snap-cap tubes (1.5 mL).

Parafilm if you will be using glass culture tubes or cuvettes for spectrophotometry.

Vortexing mixer.

Micropipettors: 20, 200, and 1000 μL.

Electric timer.

Spectrophotometer: automated plate reader, conventional spectrophotometer, or Spec 20.

Before you Come to the Laboratory

The following amounts of enzymatic activity for FNR will probably give a dA_{490}/dt for the measurements you will perform at the high end of the desirable range:

1.5 nmol min^{-1} added to an 0.2 mL reaction in a multiwell plate.
4 nmol min^{-1} added to an 0.5 mL reaction in a spectrophotometric cuvette.
25 nmol min^{-1} added to a 3 mL reaction in a Spec 20 tube.

- Calculate the volume of the concentrated FNR purified by adsorption that will be needed in each reaction. Calculate how much of a dilution of the sample you will have to make to produce a solution from which a convenient and accurate volume, for example 20 μL, can be withdrawn to add to each assay.

- The solution for the each of the assays should have the following final concentrations after the volume of diluted FNR purified by adsorption has been added:

  1. 0.5 mM INT.
  2. 70 mM NaCl.
  3. 200 mM Tris chloride, pH 9.
  4. 0, 6.25, 12.5, 25, 50, 100, or 200 μM NADPH (one assay for each concentration).
  5. The volume of diluted FNR purified by adsorption that contains the appropriate amount of enzymatic activity.

- Decide what the volume of the reactions will be for the instrument you will be using, then plan how the reaction mixes will be prepared from the available reagents. When preparing for several kinetic assays that have some components at identical concentrations and others at different concentrations, a single reaction cocktail may be prepared for the reagents that end up at the same concentrations in each assay. A particular volume of the cocktail is added to each assay, then a particular volume of the variable components, and finally, when you are ready to make a particular measurement, a particular volume of enzyme to produce the desired final volume and final concentrations and to initiate the reaction.

- The most accurate way to prepare the variable components is to take an accurate aliquot, say 200 μL, of the stock solution containing this component, in our case, 5 mM NADPH, 50 mM sodium bicarbonate, pH 9, and prepare a solution with a concentration appropriate for the highest concentration of the varied reagent—for example, a solution of 1 mM NADPH—by diluting into water. Then make serial twofold dilutions of this solution with water to get appropriate solutions; for example, in our case 500, 250, 125, 62.5, and 31.25 μM NADPH. A serial twofold dilution is made by removing half of the solution of higher concentration and mixing it with an equal volume of water to produce the lower concentration and then repeating this process again and again until solutions of each of the desired concentrations have been made. An aliquot from each of these diluted stocks—for example, an aliquot with a volume corresponding to 20% of the final volume of the reaction—will then provide each of the desired final concentrations. You should write out the volumes that you decide to use and the dilutions you decide to make before coming to the laboratory. Make sure that you make enough of the cocktail and enough of each of the dilutions of the variable component to prepare all of the assays.

- Plan what volume of a dilution of the concentrated FNR purified by adsorption you will add to each assay to bring the volume of that solution up to its final volume, all of the other concentrations up to their final values, and initiate the reaction. You should plan to add a large enough sample of diluted enzyme so that you add exactly the same amount of enzyme to each assay, but you should also plan the experiment in such a way that you can add more enzyme to the tube than you have planned to add if you have to do so. For example, you should not plan to make a stock of enzyme that is diluted by such a small factor that you cannot dilute it by an even smaller factor if the initial dilution does not add enough enzyme to the assay to see a large enough change in A_{490}.

- Mixing a set volume of each of the dilutions with a set volume of the cocktail minimizes pipetting and exposure to air, and provides more consistent results. Write out in your laboratory notebook the solutions you plan to prepare and the mixtures of those solutions you plan to make before you arrive at the laboratory.

Procedure (One Laboratory Session, 1.5 h)

- Thaw the sample of concentrated FNR purified by adsorption that you set aside for the kinetic experiments and place it on ice.

- Prepare the appropriate volumes of the seven reaction solutions you will use in the assay but do not add the enzyme at this point. After you have added the enzyme to initiate the reaction, the final volume of each of these seven solutions should be the final volume you used in your calculation in preparation for the laboratory and the final concentrations of all of the reagents should be those listed above. Protect the NADPH from atmospheric oxidation. Do not leave the stock or reaction tubes containing NADPH open any longer than necessary. Minimize the exposure of any solution containing NADPH to air by mixing the solutions gently.

- Perform the assay with 200 μM NADPH first because it will produce the greatest signal. Add the solution for this assay that does yet contain the enzyme to the tube, the cuvette, or the well. Check the A_{490} of this solution. Then add the enzyme in the volume you calculated before you came to the laboratory to bring the assay to its final volume. Start a timer immediately after adding enzyme, and take A_{490} measurements every 30 s for 3 min. If the dA_{490}/dt is 0.15 to 0.2 min^{-1}, this amount of enzyme is the appropriate one to use in the subsequent reactions. If it is not, do other assays at 200 μM NADPH with different amounts of enzyme until an appropriate rate is obtained. When you have found an appropriate amount of enzyme, perform the other assays at the lower concentrations of NADPH.

• The negative control, which is the reaction with 0.5 mM INT but no NADPH, should not show an increase in the A_{490} within the first 3 min. If it does, you should correct the other rates by subtracting this background.

Analysis of the Data and Questions to Address in Your Report

Refer to the Computational Techniques for Biochemistry for suggestions on sources for software.

1. Make two types of plots:
   a. Plot the A_{490} as a function of time for each assay. Remember in analyzing these plots that you are only interested in an accurate estimate of the initial velocity.

   b. Plot each of these initial velocities, v, as a function of the respective molar concentration of NADPH, [NADPH]. Plot [NADPH] on the x axis in units of micromolar, and the initial velocities on the y axis in units of second^{-1}. The units of second^{-1} are calculated, as before, from the observed rates for each concentration of NADPH in units of micromoles minute^{-1} milliliter^{-1} divided by the molar concentration of active sites actually in the assay. You should use the concentration of protein that you estimated for the concentrated FNR purified by adsorption in these calculations. The initial velocity in the units of second^{-1} for the concentration of NADPH of 200 μM should be about the same as you calculated previously for the catalytic constant of the enzyme. Fit Equation i.2 to the data to obtain values for its two parameters: the limiting velocity, V, and the Michaelis constant, K_{mNADPH}. The units on the limiting velocity are those used for the initial velocity, second^{-1}. The units for the Michaelis constant (Equation i.2) are those for the concentration of NADPH on the x axis. Because the data are plotted with the units of micromolar on the x axis, the value obtained for the Michaelis constant from fitting the equation to the plot will be in units of micromolar.

2. Consider the plot of the initial velocity, v, as a function of the molar concentration of NADPH, and the value of K_{mNADPH} you have calculated. Discuss whether or not the range of concentrations of NADPH that you used was appropriate to obtain an accurate value for K_{mNADPH}. Would it have been better to go to higher concentrations or have used a narrower range? If not, what would have been the disadvantages of going to higher concentrations or using a narrower range?

3. Discuss whether the K_{mNADPH} you have calculated convinces you that the enzyme you have purified is in fact FNR.

4. What does the value of K_{mNADPH} imply about the concentrations of NADPH that FNR is likely to encounter when it is participating in the cyclic electron transport around photosystem I in chloroplasts from spinach?

5. If you have already determined the K_{mNADH} of LuxG, K_{mNADH} for L-lactate dehydrogenase (LDH), or both of these kinetic constants, consider the following question. What may be concluded regarding the concentration of reduced NADPH that FNR encounters in its natural environment compared to the concentration of reduced NADH in the natural environments of the other enzyme or enzymes?

FNR EXERCISE 6

Estimation of the Size and Quaternary Structure of Native FNR Using Chromatography by Molecular Exclusion

If LDH Exercise 10, which performs the same chromatography by molecular exclusion on the same column, is on the schedule for your course, it is possible, and in fact strongly recommended, to postpone this exercise, FNR Exercise 6, until samples of both purified FNR and purified LDH are available. Both enzymes can then be run on the same column at the same time. See LDH Exercise 10 for details.

Technical Background

Chromatography by molecular exclusion (or *gel filtration*) is a method for separating molecules according to their sizes (Ackers, 1964; Andrews, 1965; Boyer, 2012; Kyte, 2007, pp 11–12, 423–430; Scopes, 1994; Segel, 1976) that uses, as a solid, stationary phase, beads of cross-linked agarose or cross-linked dextran. Chromatography is a process by which a mobile phase, in our case an aqueous buffer, moves continuously through a bed of solid phase, in our case the beaded agarose or dextran, usually situated in a column made of glass or stainless steel. As the mobile phase moves through the stationary phase, each solute in the mobile phase, in our case the molecules of protein, spends a characteristic fraction of its time immobilized on the solid phase and the rest of the time moving with the mobile phase. This is not a discontinuous process as were the elutions that you performed earlier. As the chromatography progresses, the solute is continuously stopping in the stationary phase and then entering the mobile phase and moving forward, back and forth. Because each solute spends a different amount of time stationary and the rest of the time moving with the flow and because the flow of the mobile phase through the bed is constant, each set of identical molecules emerges from the end of the stationary phase at its own characteristic elution volume, and the molecules are separated one from the other on the basis of differences in these characteristic elution volumes (Figure 1.22).

Chromatography should be distinguished from the simple adsorption and elution that you have already performed because every instant during the process of chromatography, the molecules of solute are adsorbing and eluting from the solid phase, so in a way it is as if an incredibly large number of adsorptions and elutions have been performed on the solute by the time it emerges from the bed rather than just one. Another distinguishing feature between chromatography and adsorption and elution is that small differences in the time spent immobilized can be exploited to separate the molecules one from the other. Both ion exchange and adsorption can be used, respectively, as the property that immobilizes the molecules of solute, and both chromatography by ion exchange and chromatography by adsorption are commonly used techniques to separate molecules of protein as well as chromatography by molecular exclusion.

In the case of chromatography by molecular exclusion, the molecules being separated diffuse into and out of the stationary phase, which is a beaded gel. When it is within a bead, a molecule is not moving through the chromatographic bed with the mobile phase; when it is outside of the beads, it is. The separation produced by molecular exclusion chromatography depends on the ability of the solutes to penetrate the spaces between the strands of polymer that produce the gel. The more spaces the molecule can penetrate, the more slowly it moves through the bed.

The ability of a molecule, in our case a molecule of protein in its native state, to penetrate the spaces between strands of a linear polymer, in our case linear dextran cross-linked with bisacrylamide,

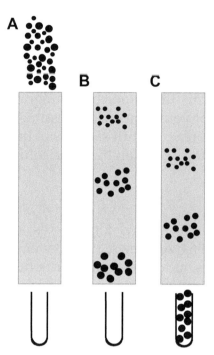

FIGURE 1.22 **Native proteins are separated by size using chromatography by molecular exclusion.** The beads of hydrophilic polymer such as cross-linked dextran or agarose are submerged in a buffered aqueous solution inside a glass or plastic cylinder. (A) A mixture of native proteins, monomeric or multimeric (black dots), is applied in a tight band of as small a volume as possible to the top of the column. The chromatographic medium (gray rectangle) consists of porous beads containing aqueous spaces of randomly distributed sizes that are more or less accessible to the sizes of the molecules of protein in the mobile phase passing through them. (B) The mobile phase is pulled by gravity through the beads. The smaller molecules of protein find more space in each bead than the larger molecules of protein find and are hindered in their movement through the bed. Molecules of protein within the resolving range of the beads, which is determined by the range of the sizes of spaces they contain, migrate at different rates. Molecules of protein and other macromolecules that cannot penetrate the beads at all emerge in the void volume before the molecules of protein that can penetrate the beads. All molecules that are small enough to gain access to all of the spaces in the beads emerge together in the included volume. (C) The eluate is collected in fractions.

is an inverse function of the surface area of that molecule. The greater the surface area of the molecule, the more spaces there are from which it is excluded. Large molecules have greater surface areas than small molecules, and other things being equal, a large molecule will be able to penetrate fewer of the spaces in the bead than a small molecule can and will therefore spend less time in the stationary phase than in the mobile phase percolating around the beads of gel. An irregularly shaped molecule, however, always has a greater surface area than a sphere of the same volume. Consequently, molecules are separated as a function of their shape as well as their volume. If the shapes of all of the molecules—for example, all of the proteins in the solution—are globular, those of the same volume will not differ greatly in their surface areas, and smaller globular molecules of lower molar mass will be retained by the gel for a greater fraction of the time than larger globular molecules of greater molar mass, and the smaller molecules of protein will be separated from the larger molecules, which pass through the column more rapidly. Globular molecules are, therefore, eluted in order of decreasing molar mass. If the molecule of protein has a highly irregular shape, such as a molecule of fibrinogen, it will appear to be larger than its actual molar mass.

In your estimation of the size of FNR, you will make the assumption that all of the molecules of protein that are used as standards, as well as FNR itself, are spheres and, consequently, that their respective rates of movement through the gel are only a function of their volume. Since the partial specific volumes of all proteins are about the same ($0.73 \text{ cm}^3 \text{ g}^{-1}$) and the average mass of an amino acid in a protein (Kyte, 2007, p. 418) is also about the same ($110 \pm 3 \text{ g mol}^{-1}$) for every protein, one can think of the characteristic *elution volume* (V_e) at which a molecule of protein emerges from the chromatographic column to be a function of its total number of amino acids, at least within the uncertainty of these three assumptions. The first two of these assumptions are always made when chromatography by molecular exclusion is used to estimate the molar mass of a molecule of protein, but molar mass is less informative than an estimate of the number of amino acids that the protein contains, and hence the number of polypeptides a molecule of protein contains in its native state.

Another way to think of the process of molecular exclusion is that the smaller molecules see a larger volume as they pass through the bed than do the larger molecules and therefore emerge from the bed at a larger elution volume. Because a molecule, however, can never see more volume in the solid phase than the volume of the bed of the column itself and because it is only diffusion into and out of the beads that is occurring, not immobilization by, for example, ion exchange or adsorption, the volume at which any solute emerges from the column must always be less than the total volume of the bed because no solute can penetrate the strands of dextran or agarose, which themselves occupy some of the volume. The total volume of the solution within the beads and the solution surrounding the beads is the *included volume* of the column. All solutes added to the top of the column emerge within or before the included volume. Molecules that are so large that they cannot penetrate any of the beads spend all of their time in the mobile phase, and their elution volume is equal to the volume in the bed that is outside of the beads. The volume of solution in the bed that is outside of the beads is the *void volume* (V_0) of the column.

Because molecules large enough that they cannot penetrate the beads at all travel in the void volume, all molecules larger than a certain size end up in the void volume and are not separated from each other. Because molecules that are small enough to penetrate all of the spaces in the bed all travel in the included volume, all molecules smaller than a certain size end up in the included volume and are not separated from each other. Only those molecules that have sizes that are small enough to penetrate the beads but not small enough to penetrate all of the space in the beads are separated from each other on the basis of their sizes.

In the case of chromatography by ion exchange or adsorption, the reason that the molecules being separated spend more or less of their time on the solid phase is that they are trapped in the layer of counter ions or physically bound to the solid phase while they are immobilized, respectively. Consequently, they usually emerge at elution volumes larger than the included volume of the column, often much larger than the included volume.

Experimental Plan

Sephacryl S-200 is a beaded form of dextran that is cross-linked by bisacrylamide. This solid phase nominally resolves globular proteins that contain 50 to 2500 amino acids. One of the folded polypeptides that constitutes FNR contains 314 amino acids. The most successful resolution of one globular protein from another will be achieved when its number of amino acids is in the logarithmic middle of the nominal range of the resolution. A chromatographic column is prepared using this solid phase as the stationary phase and an aqueous buffer as the mobile phase. A sample of the FNR that you have purified by ion exchange and adsorption and elution, and concentrated by ultrafiltration under centrifugal force, a sample of concentrated FNR purified by adsorption, is applied to the column along with a mixture of standard proteins: β-amylase, alcohol dehydrogenase, malate dehydrogenase, and cytochrome *c*. If the two procedures for purification were performed properly, the sample of FNR should contain 1 to 2 μmol min^{-1} of activity.

The standards can be identified by their absorbance at 280 nm (every protein absorbs at 280 nm), their respective enzymatic activities (β-amylase, alcohol dehydrogenase, and malate dehydrogenase) or their absorption of visible light (cytochrome *c*) (Figure 1.23). The peak of FNR can be

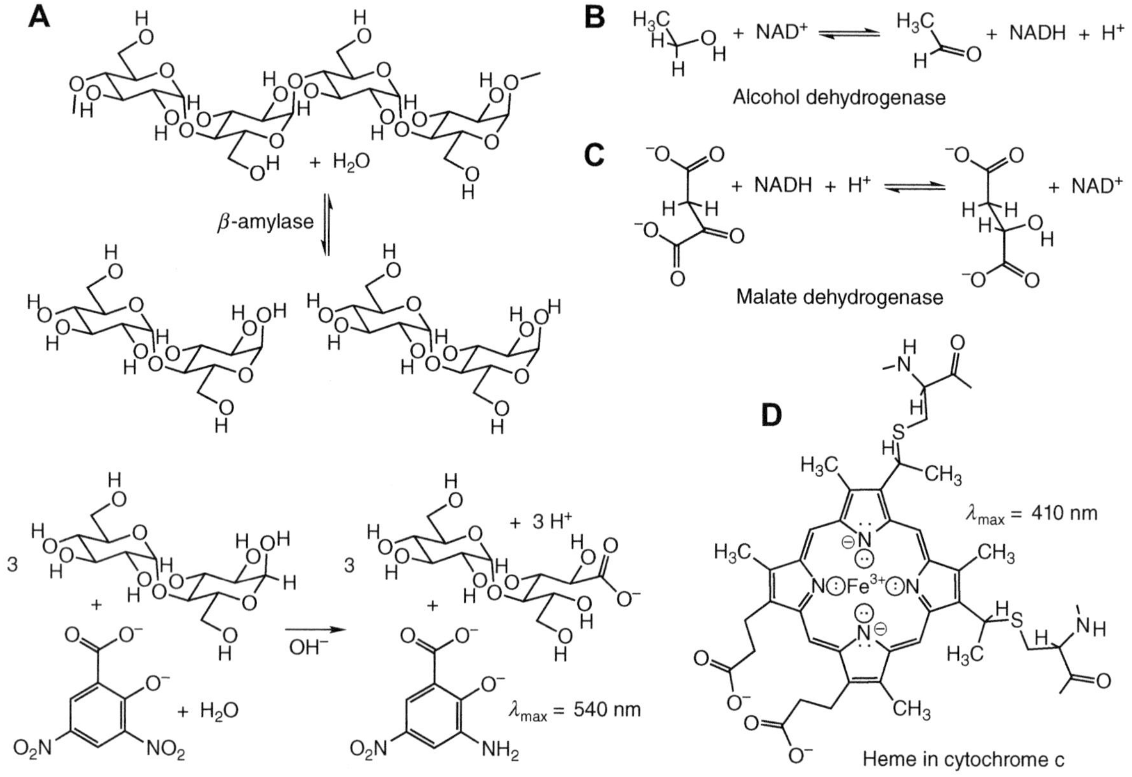

FIGURE 1.23 **Assays for the standards used in chromatography by molecular exclusion.**   (A) The enzymatic activity of β-amylase is quantified by oxidizing with 3,5-dinitrosalicylate the maltose hydrolyzed over a set interval of time from the amylose in starch. The 3,5-dinitrosalcylate is reduced by the maltose to 3-amino-5-nitrosalicylate, which absorbs at 540 nm. (B) The enzymatic activity of alcohol dehydrogenase is quantified by the increase in A_{339} as NAD⁺ is reduced to NADH by the cosubstrate ethanol. (C) The enzymatic activity of malate dehydrogenase is quantified by the decrease in A_{339} as NADH is oxidized to NAD⁺ by the cosubstrate oxaloacetate. (D) The ferroheme of oxidized cytochrome *c* absorbs at 410 nm with a ε_{410} of 106 mM⁻¹ cm⁻¹ (Margoliash and Frohwirt, 1959).

identified by its absorbance at 280 nm (from the protein), its absorbance at 450 nm (from the prosthetic FAD), or its enzymatic activity. Only the assay of the various enzymes, including FNR, for their respective enzymatic activity provides conclusive evidence for their presence. The chromatography is performed at 4 °C. All of the fractions of the eluate from the chromatographic column should be stored in the cold until assayed. Enzymatic assays, however, should be performed at room temperature.

The number of amino acids in native FNR, estimated by chromatography by molecular exclusion, and the number of amino acids in one of the polypeptides constituting FNR when it is saturated with dodecyl sulfate, estimated by electrophoresis on the gel of polyacrylamide, can then be used to estimate the number of polypeptides, and hence the number of subunits, that are contained in native FNR.

Reagents, Supplies, and Equipment

Buffered solution for the chromatography:
0.1 M sodium phosphate, pH 7.0.

Marker for the void volume of the column:
Blue dextran.

Standard proteins in the buffer for the chromatography:
β-amylase from *Ipomoea batatas* (sweet potato), Sigma-Aldrich item A8781, 4 mg mL^{-1}.

Alcohol dehydrogenase from *Saccharomyces cerevisiae* (yeast), Sigma-Aldrich item A8656, 5 mg mL^{-1}.

Malate dehydrogenase from hearts of *Sus scrofa* (pig), Sigma-Aldrich item M1567, 3 mg mL^{-1}, concentration of activity = 2100 μmol min^{-1} mL^{-1}.

Cytochrome *c* from hearts of *Equus caballus* (horse), Sigma-Aldrich item C7150, 20 mg mL^{-1} (1.7 mM).

Reagents for the assay of the enzymatic activity of FNR:
5 mM NADPH, 50 mM sodium bicarbonate buffer, pH 9.
5 mM INT, 70% ethanol, 5% Triton X-100.
1 M Tris chloride, pH 9.
5 M NaCl.

Reagents for the assay of the enzymatic activity of the standards (Figure 1.23).
1. For β-amylase:
   1% starch.
   Reagent for developing the color resulting from the reduction of 3,5-dinitrosalicylate.

2. For alcohol dehydrogenase:
   2 M ethanol.
   75 mM NAD$^+$.
   100 mM sodium diphosphate, pH 9.2.

3. For malate dehydrogenase:
   2.5 mM oxaloacetate.
   15 mM NADH.
   0.5 M sodium phosphate, pH 7.5.

Materials and equipment for chromatography:
Sephacryl S-200 in the buffered solution for the chromatography, stored at 4 °C with a low concentration of sodium azide to prevent microbial growth.

Glass chromatographic column of 2.5-cm diameter, 75 cm in height.

Mariotte bottle (Figure 1.24).

Connectors and tubing (Figures 1.24, 1.25, and 1.26).

Spirit level.

Fraction collector (Figure 1.26).

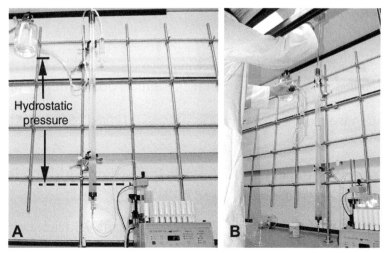

FIGURE 1.24   **Preparation of a gel filtration assembly.**   (A) Four components are connected by plastic tubing: a Mariotte bottle (upper left) that acts as reservoir for the chromatographic buffer flowing into the top of the column, the column (the one shown has diameter of 2.5 cm and a length of 75 cm), a syringe that is attached to a three-way valve on top of the column, and an automated fraction collector (lower right). The Mariotte bottle has a horizontal port near the bottom through which the buffer flows to the column. Because the bottle is stoppered at the top, the air flowing into the bottle to replace the fluid that is leaving the bottle is required to enter the bottle from the end of a long glass tube in the stopper. The glass tube ends near the bottom of the bottle. The hydrostatic pressure of the buffer flowing from the bottle through the column and to the outlet at the fraction collector is proportional to the vertical height between the end of this inlet tube in the Mariotte bottle and the outlet of the tubing connected to the drop counter of the fraction collector. Consequently, the hydrostatic pressure remains the same even though the bottle is emptying because the distance between the inlet for the air at the end of the glass tubing and the outlet of the tubing to the fraction collector remains constant. There is a two-way valve at the bottom of the column. (B) Filling the column with the Sephacryl S-200. The valve at the bottom of the column is closed, and the cap is removed from the column. Before gently pouring in the Sephacryl, about 75 mL of chromatographic buffer is poured into the column. Having buffer present before the Sephacryl is added prevents the formation of air pockets at the bottom of the bed.

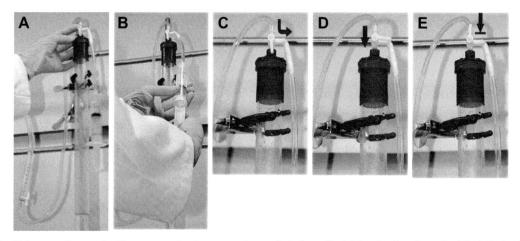

FIGURE 1.25   **Using a syringe and a three-way valve to remove trapped air from the tubing leading from the Mariotte bottle to the top of the column.**   (A) The handle of the three-way valve always points to the port that is closed. Turn the handle so that it is pointing to the top of the column to direct flow from the Mariotte bottle into the syringe. (B) Pull on the plunger of the syringe to pull all of the trapped air from the tubing, then turn the handle so that it points to the syringe to direct the flow into the column. The three valve positions: (C) handle pointing to the column, buffer flows from reservoir tubing into syringe; (D) handle pointing to the syringe, flow is from reservoir into column; (E) handle pointing to the solution flowing from the Mariotte bottle, no flow.

Materials and equipment for assay of absorbance:
If you are using a plate reader with a 200 µL sample you should use a standard plastic multiwell plate for wavelengths greater than 400 nm and a multiwell plate that is transparent to ultraviolet light for reading absorbance at 280 or 339 nm (Figure 1.13). Refer to Appendix I regarding the use of multiwell plates and the plate reader.

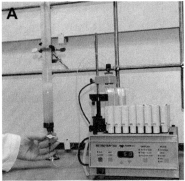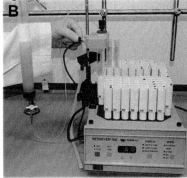

FIGURE 1.26   **Directing flow of buffer from the column into the tubes in the fraction collector.**   Plastic tubing extends from the valve at the bottom of the column to the drop counter. While you are collecting fluid in a graduated cylinder or a beaker, the arm for the drop counter can be swung out over the bench-top. When you are collecting fractions, it is swung back so that it is suspended above the collection tubes. Shut off the column before swinging it back to avoid spilling solution onto the collector. (A) The valve is opened to allow flow. (B) The position of the drop counter is carefully adjusted to direct the drops into the center of each tube that passes below it. You don't want to come in tomorrow and find your eluate flowing onto the bench-top. The test tube rack in the collector advances one position after the desired number of drops have been collected.

If you are using a conventional spectrophotometer with an 0.5 mL sample you should use a standard plastic cuvette for wavelengths greater than 400 nm and a plastic cuvette that is transparent to ultraviolet light for wavelengths of 280 or 339 nm.

If you are using a Spec 20 with a 2.5 to 3 mL sample, glass culture tubes, transparent to wavelengths greater than 339 nm, are used.

Spectrophotometer, or combination of spectrophotometers, that will assay wavelengths from 280 to 540 nm.

Precaution

Stir the Sephacryl S-200 gently and as little as possible; the solid phase easily breaks up into smaller fragments that interfere with the flow of the buffer through the chromatographic bed. Once the column has been poured, never allow it to run dry.

Procedures (Three Laboratory Sessions)

PART 1. PREPARATION OF THE CHROMATOGRAPHIC COLUMN AND ASSAY OF THE STOCKS OF STANDARDS

Setting up the chromatographic column, pouring the slurry of Sephacryl, and initiating the packing (1 h).

- Clamp the column to the stand or to the rack, check that it is perfectly vertical in all directions with a spirit level, set up the reservoir, and connect the tubing (Figures 1.24 and 1.25).

- With the valve at the outlet closed, pour about 75 mL of chromatographic buffer into the column. The presence of buffer in the column before the Sephacryl S-200 is poured in prevents the formation of air pockets at the bottom of the column.

- Using a long glass rod to guide the stream of fluid down the wall of the column, slowly and steadily add the slurry of Sephacryl S-200 until it is about 2 cm from the plastic collar at the top of the column.

- Place a 1-L beaker beneath the outlet and open the valve to start the flow. You should check the level of the bed of Sephacryl occasionally. Do not let it run dry!

- You will need to make several additions of the slurry to bring the bed of packed gel to within 5 to 10 cm from the top of the column. To make an addition, first turn off the flow. Using a pipette, remove the buffer to about 2–3 cm above the top of the bed. Use a glass rod with a flat end to stir up the top couple of centimeters of the gel bed gently. While stirring, try to make the boundary between settled gel and stirred gel as level as possible. Slowly add more slurry until you reach the top of the column, and then reopen the valve.

- Repeat this procedure as necessary. When you have reached the point where you are repeating this procedure, you and your partner can trade off taking care of the column.

Attaching the reservoir for the chromatographic buffer, final packing, and adjusting the flow rate. (Begin in the first session and complete in next laboratory session.)

- When you have added enough Sephacryl S-200, attach the reservoir and let the column flow. If necessary, use a syringe to clear air from tubing (Figure 1.25). The tubing must be free of air bubbles because they can break the siphon and stop the flow. Have the hydrostatic pressure as high as possible during packing. Let the Sephacryl pack until a volume of buffer equal to about twice the volume of the column has passed through it before shutting it off. This should take approximately two days, so you should arrange to come in to refill the reservoir the next day. Be sure to shut off the flow at the bottom of the column before refilling the reservoir.

- While the column is packing, you will have the time to assay, as described below, the standards that you will be adding to the column. You need to be certain that they are active before you add them to the column and you need to gain some practice in performing the assays.

- At the beginning of the next laboratory period, fill the reservoir, and turn on the column.

- While the rate of flow is stabilizing and the inlet tube to the Mariotte bottle is filling with air, calculate the total volume of Sephacryl in the column, V_T. Calculate the flow rate in drops minute^{-1} at which the total volume of the column would flow through in 24 h. Usually 24 drops are 1 mL, but you should put an empty tube under the outlet tube at the fraction collector, count the drops for a while, and then measure the volume that has gathered in the tube to find out how many drops are a milliliter from your particular tubing.

- Once the flow rate of the column has stabilized, adjust the height of the Mariotte bottle until the drop rate is the rate you calculated. While you are doing this you can swing the outlet tubing away from the fraction collector and collect the fluid into a beaker on the bench-top.

- Once you have established the rate of flow you want and the column has had time to come to a steady rate of flow, double check your calculations by placing a graduated cylinder under the outlet tubing that will be feeding the fraction collector and setting the timer. Collect the flow-through while you prepare the sample for the column. Before you add the sample to the column, shut off the column, stop the timer, and calculate the actual flow rate in milliliter minute^{-1} to decide whether you need to increase or decrease the height of the Mariotte flask.

Adjust the height of the Mariotte flask accordingly. There is no need to be exact here. Just get the rate of flow to within ±20% of the rate you want.

Assaying the stocks of standards for gel filtration (1.5 h).

The purpose of the following exercises is to determine the appropriate amount of each standard to load on the chromatographic column. You should assay a sample, or if necessary several samples, of a dilution of each of the standard stock solutions that will be prepared by the teaching assistant for the entire class. The teaching assistant should inform everyone as to the dilutions that he performed.

Assays for standards that lack enzymatic activity.

- For blue dextran, the marker for the void volume, the absorbance maximum is 620 nm; and for cytochrome c, it is 410 nm. For blue dextran and cytochrome c, measure the A_{620} and the A_{410}, respectively, of the diluted stocks. If either of these absorbances is greater than 0.8, dilute the sample further. Keep track of any dilutions that you need to make.

Enzymatic assay for β-amylase.

The maltose that is the product of the hydrolysis catalyzed by the β-amylase is detected by oxidizing it with 3,5-dintrosalicylate (Figure 1.23). The solution containing the 3,5-dintrosalicylate is strongly basic. Because the base denatures the enzyme, the absorbance cannot be measured continuously as a function of time. Consequently, the enzyme and the reactant are mixed together and sit at room temperature for 7 min, which is the interval that would be used for time if you were calculating rates, before the amount of product is assayed. The A_{540} observed for the sample in which the enzymatic reaction has proceeded for this interval of 7 min, however, should be proportional to the amount of enzyme in the sample, which is all that concerns the present experiment.

The concentration of the activity of the stock is not specified so several dilutions of the sample given to you by the teaching assistant should be assayed all at the same time in different tubes. Be sure to include two tubes to which no enzyme has been added to serve as background for the naturally occurring reducing ends in the starch. Because of the way this assay is performed, a sample of 0.15 mL is used in all instances, regardless of the apparatus you use to determine A_{540}.

- Mix the following two components:
  β-amylase (0.15 mL).
  1% starch (0.15 mL).

- Let the mixture sit for 7 min at room temperature, and then add 0.3 mL of the reagent for staining with 3,5-dinitrosalicylate.

- Let this mixture sit for 3 min at room temperature.

- Place each of the tubes in which you are performing the assay in a bath of boiling water for 5 min.

- Cool the mixture, and add 3 mL of water.

• Quantify the colored product by its A_{540}. An A_{540} of 0.2 to 0.7 is a reasonable yield of absorbance.

Enzymatic assay for alcohol dehydrogenase.

In the enzymatic reaction catalyzed by alcohol dehydrogenase (Figure 1.23), only NADH absorbs at 339 nm, so the rate at which the A_{339} increases is a measure of enzymatic activity. The concentration of the activity of the stock solution is not specified, so several dilutions of the sample given to you by the teaching assistant should be assayed. The final concentrations in the assay should be:

Ethanol 330 mM.
NAD+ 7.5 mM.
Sodium diphosphate, pH 9.2, 50 mM.

The fractions from the column will be in a buffer of 0.1 M sodium phosphate at pH 7.0. If too much of this is added to the assay, it will decrease the pH of the assay by too large an amount. Consequently, make the volume of the solution containing the alcohol dehydrogenase only 10% of the volume of the final assay so that the final concentration of the phosphate buffer in the assay is only 10 mM. The reaction is performed at room temperature.

• Mix the components. Add the solution of alcohol dehydrogenase last to initiate the assay, and immediately begin to measure A_{339} at intervals of 30 s for 3 min. As with the assay for FNR, a dA_{339}/dt between 0.05 and 0.2 min^{-1} is a reasonable rate.

Enzymatic assay for malate dehydrogenase.

In the enzymatic reaction catalyzed by malate dehydrogenase (Figure 1.23), only NADH absorbs at 339 nm, so the rate at which the A_{339} decreases is a measure of enzymatic activity. The concentration of the activity of the stock solution is not specified, so several dilutions of the sample given to you by the teaching assistant should be assayed. The final concentrations in the assay should be:

Oxaloacetate 0.25 mM.
NADH 0.15 mM.
Sodium phosphate, pH 7.5, 100 mM.

The fractions from the column will be in a buffer of 0.1 M sodium phosphate at pH 7.0, which is close enough to the pH of the assay to have no effect. Consequently, make the volume of the solution containing the malate dehydrogenase as large a fraction of the assay as possible so that the malate dehydrogenase that you have to add to the column is not wasted. The reaction is performed at room temperature.

• Mix the components. Add the solution of malate dehydrogenase last to initiate the assay, and immediately begin to measure A_{339} at intervals of 30 s for 3 min. As with the assay for FNR, a dA_{339}/dt between −0.05 and −0.2 min^{-1} is a reasonable rate. The rate is negative in this case rather than positive as it was for the assay of FNR activity because the absorbance is decreasing rather than increasing.

Again, for all of these assays, the actual enzymatic activities are inconsequential. You only performed these assays to discover how much of each standard to add to the chromatography to get signals that are large enough when you are assaying fractions from the column.

PART 2. PREPARING THE MIXTURE OF PROTEINS FOR CHROMATOGRAPHY, LOADING THEM ON THE COLUMN, AND COLLECTING FRACTIONS OF ELUATE

Preparing the mixture of standards and FNR purified by adsorption for the column (30 min).

The peaks in the concentrations of each of the standard proteins and the peak of the concentration of FNR emerging from the column are usually about 20 mL wide. It is as if the chromatography were diluting the samples you add into a final volume of 20 mL. You can now calculate what volumes of the stock solutions of blue dextran and cytochrome *c* you have to add to produce reasonable absorbances for these molecules in 20 mL of the chromatographic buffer. You can also calculate from the enzymatic assays you just performed, and the dilutions that were performed by the teaching assistant and yourself, how much of each undiluted standard you would have to add to 20 mL to obtain a convincing assay of each of the standards if the volumes you are adding to each assay were being directly removed from that 20 mL.

After you have calculated how much of each standard to add to the chromatographic column, check with your neighbors and come to an agreement on how much of each you should add to the columns and then check your calculations with the teaching assistant. The reason for coming to general agreement, and presumably the correct values, is to avoid using too much of a standard and wasting it, or too little and not being able to find it.

Remove the sample containing approximately 1.6 μmol min^{-1} of concentrated FNR purified by adsorption from the freezer. This sample is the one you set aside after concentrating the pool of enzymatic activity. Thaw and keep the sample on ice.

Add the appropriate volumes of each of the standards to this sample containing the FNR. This mixture is the sample that you will add to the column and that the chromatography will separate.

Add glycerol to 7.5% (75 μL of neat glycerol mL^{-1}) to increase the density of the mixture. Assume that each drop of glycerol from a pipette is 35 μL. Mix thoroughly to dissolve the glycerol. You should spin the mixture in the microcentrifuge for a few seconds to get rid of bubbles in the mix, which would bedevil the loading of the column.

Applying the sample to the column and setting up the fraction collector (2 h).

Shut off the flow at the bottom of the column, and then shut off the flow from the reservoir to the column at the top by turning the valve so the handle points to the tube from the reservoir. Refill the reservoir. Remove the fitting at the top of the column.

Take the sample for the column up into a pipette so that there is no airspace at its tip. Carefully lower the tip of the pipette below the level of the fluid on top of the column and empty it slowly so that the sample slowly sinks and evenly spreads over the top of the bed of Sephacryl below the buffer. It should form a tight band on the top of the Sephacryl. The smaller and tighter the sample; the better the resolution of the chromatography.

- Empty the graduated cylinder under the outlet of the tubing that you have been using to adjust the rate of flow, place it back under the outlet, connect the reservoir and turn on the flow. You will need to measure the volume of buffer flowing out of the column as explained below. You may collect the buffer in a beaker if that is more convenient, but you then must carefully measure the volume collected in the beaker in a graduated cylinder.

- Wait until bubbles are emerging from the inlet to the Mariotte flask, and then note the volume in the graduated cylinder or the beaker and set the timer. The graduated cylinder or the beaker is supposed to collect all of the fluid flowing from the column after you have loaded the sample and before you hook up the fraction collector. This volume will be needed to calculate the elution volume of each standard and the void volume as explained later.

- After 30 min, calculate the flow rate of the column. If the total volume of the column is not going to emerge in 24 ± 4 h, adjust the flow rate accordingly.

Collecting fractions of eluate (overnight step).

- Set the fraction collector to collect 60 drops in each fraction. This number of drops will make each fraction around 2.5 mL. Tubes for the collection of the fractions should be numbered before you load them into the collector (Figure 1.26).

- Set up the fraction collector, and begin collecting fractions.

- Measure and record the volume of buffer that was collected in the graduated cylinder or the beaker from the time the sample was loaded until the drops begin falling into the tubes.

- Make certain you have the collector racks and tubes set up so that the machine will stop advancing the racks after the last tube has collected a fraction. If the racks continue to advance after the last tube has collected, there will be a second fraction added to each tube. Deconvoluting this problem can be painful.

- Watch the fraction collector for at least 10 tubes. Be certain that tubing or wires are not going to catch on the test tubes and that the drops are falling directly into the tubes.

- You or your partner must come by the laboratory both in the morning and in the late afternoon of the next day to check the column.

- Both before you leave today and when you come in tomorrow, check again that the flow rate is sufficiently fast that the total volume of the column is going to emerge in around 24 h by checking the volume of the fractions and how many fractions have been collected. Adjust the flow rate if necessary.

- Check the buffer in the Mariotte bottle, and add more if necessary. Remember to stop the column while you are filling the bottle.

- When checking the column, remove full tubes from the collector and insert fresh tubes as necessary. Keep the numbered tubes that you remove from the fraction collector in proper order as you remove them from the fraction collector. Keep all fractions at 4 °C until they are assayed.

After you have completed the chromatography, stop the flow through the column by closing the valve at the bottom, and save the column for LDH Exercise 10 if that is on your schedule and you will be doing that chromatography separately from the one you are now doing.

PART 3. ASSAYING THE FRACTIONS OF ELUATE FOR THE PRESENCE OF STANDARDS AND FOR ENZYMATIC ACTIVITY OF FNR

Elution volumes. The elution volume for a particular protein (V_e) is the volume of buffer that flows through the column between the time at which the mixture of enzymes, proteins, and the marker entered the bed of the column at the top and the time at which the middle of the peak of absorbance or enzymatic activity for that protein exits the tubing feeding the fraction collector. Consequently, the elution volume of a standard or of the FNR is the sum of (1) the volume of chromatographic solution that flowed out of the column between the time when the proteins and blue dextran were loaded on the top of the column and when the first fraction was collected and (2) the sum of the volumes of the all of the fractions that were collected before the peak of absorbance or enzymatic activity for the particular component eluted.

The fractions should all be the same volume if the collector functioned properly. If this is the case, you only need to measure the volume of several fractions chosen at random throughout the rack of fractions. If the volumes are all about the same volume with only random fluctuations, take the average of the volume to be the volume in each fraction. If there is a continuous decrease in volume or increase in volume through the rack of tubes, the size of the drops decreased or increased with time during the chromatography, and you must include this decrease or increase in volume over time in the calculation of the elution volumes. If the fraction collector misbehaved, some of the volumes of the fractions may vary. Measure and record the volumes of fractions that are atypical. They can be found easily by scanning the rack of tubes visually. You must have accurate volumes for the fractions because you will use the individual volumes in the tubes to calculate by summation the elution volumes, V_e, of each protein.

Void volume. Blue dextran is a collection of molecules of dextran to which a blue dye is covalently attached. Most of the molecules of dextran are so large and so irregular in shape that they cannot penetrate the beads at all, but there may be some of the blue dextran molecules that are smaller and elute at a volume greater than the void volume. The volume at which the first peak of A_{620} from the blue dextran emerges from the tubing is the void volume, V_0, of the column plus any volume between the bottom of the Sephacryl and the end of the tubing feeding the fraction collector.

Assaying the fractions (4 h).

It is easiest to identify the peaks of cytochrome *c* and *β*-amylase first. The cytochrome *c* is colored and can be located easily. The *β*-amylase has an elution volume only a little greater than the blue dextran, so you can search in this area to find it. The advantage of looking for *β*-amylase is that multiple assays can be done at the same time. Be certain to do a negative control for the *β*-amylase assay, that is a reaction with no sample from a fraction. A molecule of alcohol dehydrogenase has about three-quarters the size of a molecule of *β*-amylase, so its elution volume should be near that of *β*-amylase but further along. A molecule of malate dehydrogenase has about a half the volume of a molecule of alcohol dehydrogenase so its elution volume should be between that of alcohol dehydrogenase and cytochrome *c*. Use the assays you have already practiced to find these three standard enzymes.

* Assay samples from the fractions for enzymatic activity of FNR as you have done before. You can find FNR by looking for the yellow color of its prosthetic FAD, but only an assay of enzymatic activity is conclusive.

* Almost all proteins have a maximum of absorbance in the ultraviolet at 280 nm. Because the A_{280} arises almost entirely from the absorbance of the tryptophans a molecule of protein contains, and tryptophan is the most variable of the amino acids in the composition of proteins, the A_{280} for every milligram of protein (milliliter)$^{-1}$ will vary widely. Usually, however, the A_{280} of a solution of protein with a concentration of 1 mg mL^{-1} will be between 0.5 and 2.0.

* You should be able to locate the various proteins in the fractions from the column by their A_{280}. Measure the A_{280} of the fractions containing the various enzymes to obtain a more accurate determination of where the peaks of the proteins occur in the fractions.

Analysis of the Data and Questions to Address in Your Report

1. Plotting raw data for the fractions from the column:
  a. Manual plotting: On the same sheet of graph paper, plot for each protein all of its signals (absorbance or enzymatic activity or both) detected in each fraction. Volume should be the units for the abscissa of the plot. Place the points for each of these values for each of the proteins at the position on the plot for the volume of buffer that had flowed out of the column between the time the sample entered the top and the time at which the particular fraction was being collected. Do not forget to add in the volume that passed through the column after the sample entered and before you started the fraction collector. You may use different colors for the points and the curve for each protein to keep them separate. Draw a smooth, symmetrical, Gaussian bell-shaped curve through all of the data for each protein. The position of the elution volume, V_e, for each protein is at the maximum of this curve.

  b. Alternatively: Use spreadsheet software to produce the same graph.

2. Go to the UniProt database at http://www.uniprot.org to find the length of the mature constituent polypeptide and the number of subunits in the native structure of each of the standards. This database lists the mature length in addition to the length translated. Make sure you are using the mature length. Use these two facts, the mature length and the number of subunits, to calculate the number of amino acids in each of the standard proteins in its native oligomeric structure contains.

3. The *void volume* (V_0) of the column is the volume of the space that is between the beads and outside of them. Since the blue dextran cannot enter the beads, it remains between them and outside of them and emerges from the tubing feeding the fraction collector at a volume equal to the void volume of the column plus the volume of the tubing and the volume of the fluid between the bottom of the Sephacryl and the beginning of the tubing. The total volume of the column, V_T, that you measured after the bed had settled, is the sum of the void volume and the volume occupied by the beads of Sephacryl. The volume in the bed occupied by the beads (V_b) is the total volume, V_T, of the bed minus the void volume, V_0. By ignoring the volume of the tubing and the volume of the fluid between the bottom of the Sephacryl and the beginning of the tubing, the void volume, V_0, can be approximated by the elution volume of the blue dextran, and the volume in the bed that is occupied by the beads, V_b, is approximately the total volume of the column minus the elution volume of the blue dextran. The fraction ($\alpha_{acc,i}$) of the volume in the beads that was accessible to a particular protein i is the elution volume of protein i, $V_{e,i}$, minus the elution volume of the blue dextran, V_0, divided by the volume occupied by the beads, V_b (Equation 1.4). Convince yourself that this is correct. The term *partition coefficient* is sometimes used for $\alpha_{acc,i}$ (Ninfa et al., 2010).

$$\alpha_{acc,i} = \frac{V_{e,i} - V_0}{V_T - V_0} = \frac{V_{e,i} - V_0}{V_b} \tag{1.4}$$

The fraction of the volume inside the beads that is accessible to a molecule of protein is a function of only its surface area, which is, within the approximations already discussed, proportional to the number of amino acids it contains raised to the 2/3 power. Plot the negative of the natural logarithms of α_{acc} for the proteins used as standards as a function of their contents of amino acids, aa_i, raised to the 2/3 power (Kyte, 2007, pp 424–425). Use this standard curve to estimate by interpolation the number of amino acids in native FNR from its elution volume. If you are more traditional and prefer to estimate the molar mass of FNR, you may plot the negative of the natural logarithms of α_{acc} for each of the proteins used as standards as a function of their molar masses raised to the 2/3 power.

Hint: The values of α_{acc} will all be between 0 and 1. Thus, you will be plotting the negatives of natural logarithms of numbers between 0 and 1. In this case, a semilog plot generated by spreadsheet software with a logarithmic axis may be difficult to interpret. Thus it is best to calculate the negatives of the natural logarithms with the spreadsheet first and then plot these negative logarithms on a linear scale.

4. Based on your results, does FNR from *S. oleracea* appear to be a monomer or a dimer? Discuss any published data that you can find that is relevant to this issue and compare this published data to your findings.

5. The Experimental Design section presented later in this book includes an exercise addressing association of FNR with the membrane of the thylakoid. Even if this exercise is not on your course schedule, read the instructions and consult the appropriate literature on this subject (Lintala et al., 2007; Zanetti and Arosio, 1980; Zhang et al., 2001). Discuss how the results of the molecular exclusion experiment would affect interpretation of experiments addressing localization of FNR on the thylakoid.

6. Discuss whether the exercises studying FNR in this section allow you to make any conclusion regarding whether or not the enzyme is localized on the thylakoid.

7. It has been reported that newly synthesized FNR enters the chloroplast in an *apo* form that does not yet have its prosthetic FAD (Onda and Hase, 2004). This report suggests that association of FAD with the enzyme, which creates the *holo* form, correlates with thylakoid localization. In light of the possibilities discussed in this report, discuss whether or not it is possible that some of the FNR you detected on the chromatogram from the molecular exclusion lacked FAD.

FNR EXERCISE 7

Detection of FNR in Extracts from Chloroplasts of Spinach by Immunoblotting

Technical Background
DEFINITIONS OF TERMS USED IN IMMUNOCHEMISTRY

See Kyte (2007, pp 555–572).

An *antigen* is a macromolecule, a complex of macromolecules, or a microorganism such as a virus or a bacterium that elicits an immune response when the immune system of an intact living animal is exposed to it in some way. An antigen must be larger than about 15,000 g mol^{-1} or it must have been converted into some sort of complex larger than this threshold to elicit an immune response, because the immune system has evolved to eliminate foreign macromolecules or viruses, not small molecules. All of the antigens used to elicit immune responses for the purpose of the experiments you will be performing are large enough molecules of protein to be effective antigens.

An *epitope* is a specific location on the surface of an antigen that has a particular molecular structure and that is recognized by a particular antibody or a set of specific antibodies that the epitope elicits during the immune response. Antigens usually have several epitopes on their surfaces, each of which elicits its own set of immunoglobulins.

Immunoglobulins are molecules of protein in an animal. They are present in the blood, in the lymph, on the surface of lymphocytes, and in various excretions. Each immunoglobulin of a particular type—for example, immunoglobulin G (IgG), IgM, or IgE—has a similar structure even though they were probably elicited by different antigens and bind to only the antigen that elicited them. The majority if not all of the immunoglobulins that you will be using in these experiments are immunoglobulins G.

An *antibody* is an immunoglobulin generated by the immune system of an animal in response to an antigen. An antibody is an immunoglobulin that binds specifically to that antigen.

Antiserum is serum obtained from the drawn blood of an animal the immune system of which has been exposed to a particular antigen. An antiserum contains some immunoglobulins that are specific to the antigen of interest and that bind to epitopes on its surface, but most of the immunoglobulins in the antiserum do not recognize the antigen of interest. Mammals have several different types of immunoglobulins in their serum. Within a given type of immunoglobulin in the serum, there are a large array of individual immunoglobulins each specific to an antigen to which the animal has been exposed during its lifetime. The immune response to the antigen of interest for the experiments you will be performing generates antibodies specific to that antigen in the serum of that animal in addition to the immunoglobulins that were already there before and that are specific to antigens to which the animal had already been exposed at an earlier time. Usually, because the animal was exposed to the antigen of interest more recently than the others, because the antigen of interest was injected into the animal in a conjugate that heightens the immune response, and because the antigen of interest is injected into the animal initially and then as a boost after a period of about six weeks, the antibodies that bind to the antigen of interest are in a higher concentration in the serum than each of those that is specific to an earlier antigen.

A *B-cell* is a lymphocyte that circulates in the blood and the lymph and that is responsible for producing and secreting a particular immunoglobulin. All of the immunoglobulins that are secreted by one B-cell are identical in sequence and affinity for an epitope. One B-cell, one immunoglobulin. Only a handful of the millions of B-cells in an animal happen to make an immunoglobulin that can bind to a particular epitope. When the immune system is exposed to an antigen, that handful of B-cells that happen to produce an immunoglobulin that can bind to any of the epitopes on that antigen are stimulated to start dividing and secrete their individual immunoglobulins. All of the progeny of one of these different B-cells that are stimulated by the presence of the antigen to divide secrete the same immunoglobulin with the same sequence of amino acids as their parental cell, and each is considered to be a clone of the parental cell. All of these progeny together form a clone, even though they disperse separately and at random throughout the blood and lymph of the animal while they continue to secrete significant amounts of their own particular immunoglobulin.

A *polyclonal* set of antibodies is a mixture of immunoglobulins, all specific to one antigen. Each member of the polyclonal set has its own sequence of amino acids, its own affinity for its particular epitope, and is produced by B-cells of the same clone. Each member of the set is different from the other members of the set in sequence and affinity for a particular epitope. Antiserum taken directly from an animal that has been exposed to a particular antigen almost always contains a polyclonal set of antibodies that recognizes one or more epitopes on that antigen by binding to it.

Affinity adsorption is a method that can be used to purify the polyclonal set of immunoglobulins that is specific to the antigen that was used to elicit the immune response in the first place from all of the other immunoglobulins in the serum that do not recognize that antigen. It is the same procedure as the one you have already used to purify FNR, or if you have already done so, LDH. The antigen is covalently attached to a solid support. The antiserum is mixed with the adsorbent, and the antibodies in the antiserum that bind to epitopes on the antigen bind to the molecules of antigen and become adsorbed to the solid phase. After the solid phase has been thoroughly rinsed, the antibodies might be removed by increasing the ionic strength, which is a mild treatment, but in most cases they are temporarily unfolded with acid or another denaturant just enough that they dissociate from the antigen, which is a harsher treatment. Because the antigen is in a different state from what it was when it was injected into the animal, because the affinities of some of the antibodies recognizing that antigen are not high enough to remain bound to it during the rinse, or because the acid or the denaturant sometimes doesn't release all of the antibodies from the solid phase, affinity adsorption usually selects only a subset of the polyclonal set of antibodies in the serum. It does select, however, those antibodies that bind tightly enough to the antigen to survive rinsing and that can be released if necessary by mild treatment with acid or another denaturant. After purification, the term antibody is used rather than antiserum because there is no longer any serum associated with the antibodies.

The *titer* of an antiserum is a measure of the concentration of antibodies in the serum that recognize a particular antigen or a particular epitope with an affinity for that antigen or that epitope high enough that they cling to it even while the rest of the serum is being washed away. The titer of an antiserum can be quantified by an enzyme-linked immunosorbent assay (ELISA) (Aviva Systems Biology, 2012; Engvall, 1980). The titer measured by an ELISA is usually expressed in relative units rather than molar concentration of specific, high-affinity immunoglobulins. For example, several bleeds are usually performed from a host animal. If a 1:10,000 dilution of the third bleed gives as strong a signal in an ELISA as a 1:5000 dilution of the second bleed, the third bleed has a twofold higher titer than the second.

A *hybridoma* is a cell that can be grown in culture and that produces immunoglobulins that all have the same sequence of amino acids and consequently the same affinity for only one epitope on an antigen that has been chosen by the investigator (Kohler and Milstein, 1976). Such a cell is created in the following way. A mouse is injected with antigen. When the level of the response of the immune system of the mouse to the antigen is at its peak, indicating that there are high concentrations of B-cells secreting antibodies specific for that antigen, the animal is killed and its spleen removed. The spleen in a mammal contains high concentrations of the lymphocytes in that animal. The B-cells from the spleen are dispersed and mixed with myeloma cells, which are an immortal line of cells that can be multiplied to high levels by culturing. A hybridoma is formed when an individual B-cell fuses with an individual myeloma cell. Hundreds of thousands of hybridomas are produced in a test tube during this fusion. A hybridoma secreting a particular antibody that recognizes an epitope on the antigen of interest can be selected from this heterogeneous mixture. When a single hybridoma, which is a single cell, is then grown up in culture, all of the progeny secrete the same immunoglobulin, all the molecules of which have the same sequence of amino acids and affinity for only one of the epitopes on the antigen that was initially used to stimulate the immune system.

A *monoclonal antibody* is an antibody obtained from the medium of a culture of a hybridoma (Chakhtoura and Abdelnoor, 2010). If this one antibody has not been purified sufficiently from the medium of the culture, the preparation of the monoclonal antibody may contain a variety of proteins, but it will contain only one immunoglobulin that recognizes an epitope on the antigen of interest.

The *primary antibody* in an immunochemical experiment is the immunoglobulin that binds with high specificity to the antigen of interest that was used to stimulate the immune system in the animal from which it was obtained.

The *secondary antibody* in an immunochemical experiment is an antibody that binds to the primary antibody. An immunoglobulin is a molecule of protein and, as such, immunoglobulins can be used as antigens to elicit a secondary antibody in the same way that the primary antibody was elicited. The secondary antibody recognizes epitopes on the surface of the primary antibody and binds to them with high affinity. An animal of a different species than the animal in which the primary antibody was produced must be used to produce a secondary antibody because an animal will not make antibodies against its own proteins. For example, if the primary antibody was raised in a rabbit, an appropriate secondary antibody would be one elicited in a goat, and the product would be caprine anti-rabbit antibody. The antibodies in this serum raised in a goat are produced commercially and in such a way that they recognize epitopes on any immunoglobulin of a particular type, such as immunoglobulin G, from a rabbit that are in regions of that immunoglobulin distinct from the site at which the immunoglobulin from the rabbit binds to the primary antigen. This is done so that every immunoglobulin of that type, regardless of the epitope it recognizes, is bound by the secondary antibody and so that the secondary antibody will bind with high affinity to the complex between the primary antibody and the primary antigen, even though the site on the primary antibody bound to the epitope of the antigen is inaccessible in the complex.

THEORY AND PRACTICE OF THE IMMUNOBLOT TECHNIQUE

SIGNIFICANCE OF THE TECHNIQUE   When a solution of proteins is submitted to electrophoresis after the polypeptides it contains have been unfolded and coated with dodecyl sulfate, the resulting stained gel of polyacrylamide is a catalogue of the number, relative amounts, and the lengths of the polypeptides that solution contains. If the length of the polypeptide constituting the protein being studied is already known, the detection by nonspecific staining such as Coomassie Brilliant Blue of a band at the position on the gel for that length is consistent with the polypeptide represented by that band being the polypeptide of interest, but it does not prove that the polypeptide represented by the band is the one constituting the protein of interest. An immunoblot provides conclusive evidence that the polypeptide at that position is the polypeptide of interest (Burnette, 1981; Gershoni and Palade, 1983; Kurien and Scofield, 2009; Towbin et al., 1979).

ELECTROPHORETIC SEPARATION OF POLYPEPTIDES AND TRANSFER TO NITROCELLULOSE   Dodecyl sulfate is added to a sample from the solution of proteins to saturate every polypeptide it contains, and the resulting complexes between the unfolded polypeptides and ions of dodecyl sulfate are separated vertically by electrophoresis on a gel of polyacrylamide. The polypeptides are then transferred laterally to a sheet of nitrocellulose (Figures 1.27 and 1.29) to create a replica of the gel on

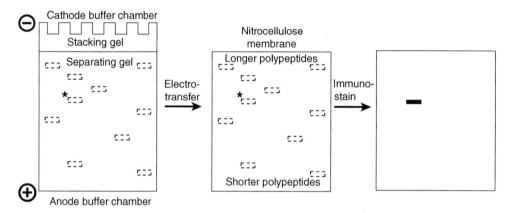

FIGURE 1.27   **Summary of the technique of immunoblotting.**   The polypeptides saturated with dodecyl sulfate are separated vertically according to their length by electrophoresis on a gel of polyacrylamide. The complexes between the polypeptides and dodecyl sulfate are then transferred laterally, again by electrophoresis, onto a sheet of nitrocellulose. The polypeptide of interest (indicated by an asterisk) is not yet visible. The sheet is then immunostained, and only the protein recognized by the primary antibody, which was elicited by the purified polypeptide of interest as an antigen, is detected.

the sheet of nitrocellulose. Nitrocellulose has the useful property of binding polypeptides tightly and nonspecifically, so the sheet can be washed to remove the dodecyl sulfate and then treated with various solutions without washing away the polypeptides. This property of nitrocellulose is used in ways other than immunostaining. Because Coomassie Brilliant Blue, which stains all polypeptides, is not used for detecting the polypeptides transferred to the nitrocellulose, covalently stained polypeptides of known length are used to calibrate the electrophoretic separation.

IMMUNOSTAINING   Rather than detecting all the polypeptides on the sheet of nitrocellulose as a nonspecific stain does, immunostaining detects only a polypeptide that has been used as an antigen to elicit antibodies that bind only to that polypeptide (Figures 1.27 and 1.28). The method, instead of staining the polypeptide of interest, stains the specific antibodies that are bound only to the polypeptide of interest. The immunostain uses a secondary antibody to which a peroxidase has been covalently attached. The secondary antibody binds to the primary antibody that is bound to the polypeptide of interest. The peroxidase is then used to produce enzymatically an insoluble colored product at the position on the nitrocellulose of the polypeptide of interest.

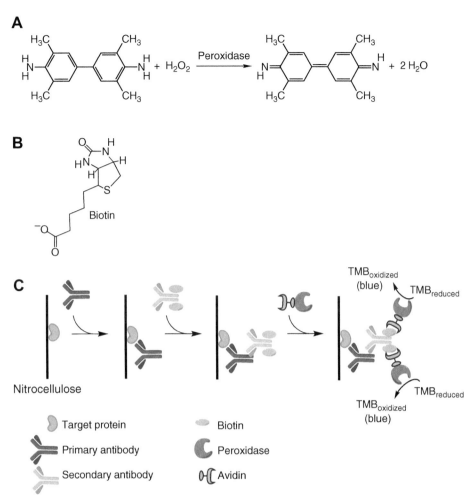

FIGURE 1.28 **Immunostaining with peroxidase to detect a particular polypeptide on a sheet of nitrocellulose.** (A) The two reactants for peroxidase that produce the staining are hydrogen peroxide and the chromogen 3,3′,5,5′-tetramethylbenzidine Peroxidase catalyzes an oxidation–reduction in which the equivalent of two hydrogen atoms are removed from 3,3′,5,5′-tetramethylbenzidine to produce the oxidized product 3,3′,5,5′-tetramethylbenzidine diimine while hydrogen peroxide is reduced to water. 3,3′,5,5′-Tetramethylbenzidine diimine forms a blue precipitate on the sheet of nitrocellulose. (B) The structure of biotin (vitamin H). It is usually used as a prosthetic group in the active sites of enzymes that transfer carbon dioxide to enolates and enamine to carboxylate them. Biotin is usually bound covalently in the active sites of these enzymes through an amide between a lysyl side chain from the protein and the carboxyl group of the biotin (Elia, 2010). (C) Schematic drawing of the process of immunostaining. Avidin is a protein from the whites of eggs that stores biotin for the embryo. The high affinity of avidin for biotin is exploited in forming the complex that attaches peroxidase ultimately to the polypeptide being stained (Pugliese et al., 1993). The primary antibody is usually a rabbit polyclonal antibody while the secondary antibody is typically an antibody raised in goats against the constant region of rabbit immunoglobulin G so that it recognizes all immunoglobulins G (Harlow and Lane, 1999). In this example, the secondary antibody is covalently conjugated to biotin, and the peroxidase is conjugated to avidin. The fact that several secondary antibodies usually bind to the primary antibody (not shown in the drawing for the sake of clarity) and the fact that there are several biotins attached to each secondary antibody cause multiple molecules of peroxidase to be bound to each primary antibody, which in turn is bound to the polypeptide adsorbed to the sheet of nitrocellulose. Consequently, the signal is amplified.

By using peroxidase, a molecule of which is attached to each of the several secondary antibodies attached to each primary antibody and which produces the stain enzymatically, visible color can be produced at a site on the sheet of nitrocellulose containing as little as 0.1 ng of the polypeptide recognized by the primary antibody (Towbin et al., 1979).

MINIMIZING NONSPECIFIC BACKGROUND Immunoglobulins are proteins and nitrocellulose binds all proteins tightly and nonspecifically. For this technique to be effective, the proteins transferred from the gel must adhere tightly to the membrane, but the antibodies used for detection must not be allowed to adhere to the sheet of nitrocellulose. A clean result is obtained only if the immunoglobulins associate specifically at the location of the target protein on the sheet of nitrocellulose and nowhere else. To prevent nonspecific association of immunoglobulins with the nitrocellulose, the sheet is saturated with a blocking reagent, which is usually an inexpensive protein that is not recognized by any of the immunoglobulins that will be applied. The sheet of nitrocellulose is soaked in a solution with a high concentration of blocking reagent just before adding the antiserum or the solution of antibodies, and the antiserum or the solution of antibodies is prepared in a solution containing the blocking reagent so it is present during the association of the antibodies with the antigen. Examples of blocking reagents are bovine serum albumin or the solids from nonfat milk, which is an inexpensive solution with a high concentration of protein. The nonionic detergent Tween-20, however, has also been found to be an effective blocking reagent and is used in the procedure that you will be following.

Potential Hazard

Although tests that have been performed on the substrate for the peroxidase, 3,3',5,5'-tetramethylbenzidine (Sigma-Aldrich company), have not revealed any toxicity, it would be best to wear gloves when handling solutions of this reagent because molecules related to 3,3',5,5'-tetramethylbenzidine are known to be toxic.

Reagents, Supplies, and Equipment

Samples of the proteins to be assayed:

Pool from DEAE: The 0.1% sample set aside after concentrating the pool and before submitting it to adsorption and elution.

Concentrated FNR purified by adsorption: The sample of 1 μg, set aside after concentrating the pool and before performing the electrophoresis and the chromatography by molecular exclusion.

Antiserum and antibodies.

Primary antiserum:

Rabbit antiserum elicited when FNR from *Arabidopsis thaliana* was used as the antigen. Although this antiserum was raised against FNR from *A. thaliana*, it is known to cross-react with the polypeptide of FNR from *S. oleracea* on immunoblots.

Secondary antibody:

Antibody elicited in goats against the constant region of rabbit immunoglobulin G. This antibody has been purified by affinity adsorption to immobilized rabbit immunoglobulin G, and biotin has been covalently conjugated to it.

Reagents for staining:

Peroxidase covalently conjugated to avidin: Extravidin-Peroxidase (egg white), Sigma-Aldrich item E2886.

Solution containing the chromogenic substrate, 3,3',5,5'-tetramethylbenzidine:

Sigma-Aldrich item T0565. Although Sigma-Aldrich describes it as a "mildly acidic solution" they do not describe all the components of the solution. It must, however, contain hydrogen peroxide, the other reactant for peroxidase, because hydrogen peroxide does not have to be added to produce the deposition of 3,3',5,5'-tetramethylbenzidine diimine.

Buffered solution for transferring denatured polypeptides from the gel to the sheet of nitrocellulose:
   192 mM glycine, 20% methanol, 25 mM Tris, pH 8.6 (Towbin et al., 1979).

Buffered solutions for soakings of the membrane and washes:
   Phosphate-buffered saline: 150 mM NaCl, 10 mM sodium phosphate, pH 7.5.

   Phosphate-buffered saline with Tween-20: 0.05% (v/v) Tween-20 in phosphate-buffered saline. Tween-20 is a nonionic detergent. The name Tween is a trademark of Uniqema Americas LLC. The name used by the Chemical Abstract Services (CAS) is poly(oxy-1,2-ethanediyl) sorbitan monododecanoate. Alternative names are polyethylene glycol sorbitan monolaurate or polyoxyethylenesorbitan monolaurate.

Sheet of porous nitrocellulose for the adsorption of polypeptides:
   Immobilon-NC HAHY, Sigma-Aldrich item N6645. Composed of nitrocellulose and some cellulose acetate with pores of 0.45 μm (Immobilon is a trademark of Merck KGaA).

Miscellaneous materials:
   Blunt-ended forceps for handling filters.

   Rectangular plastic containers for soakings and washes that should be only slightly larger than the sheet of nitrocellulose so that volumes are kept to a minimum.

Equipment for the electrophoretic transfer of proteins from polyacrylamide to nitrocellulose:
   An example of the type of apparatus required is shown in Figure 1.29.

Electrophoresis and photographic imaging:
   Use the reagents, supplies, and equipment that were used before for electrophoresis on gels of polyacrylamide cast in solutions of dodecyl sulfate, with the following exceptions: (1) the Simply Blue stain and microwave oven are not needed, (2) prestained markers of known length (ColorBurst, 8–210 kDa, Sigma-Aldrich item C1992) are used rather than those mentioned in the previous exercise.

Precaution

Acrylamide is a neurotoxin that can be adsorbed through the skin: wear gloves and use pipette bulbs. Never pour acrylamide down the drain. Polymerized acrylamide is much less toxic. Polymerize any unused acrylamide and then toss the solid into the trash.

Before you Come to the Laboratory

The calculations you performed for the exercise in which you separated polypeptides coated with dodecyl sulfate by electrophoresis should be a guide for the calculations you must perform before this exercise in immunoblotting. In the previous exercise, proteins were detected with Coomassie Brilliant Blue, which has a limit for detection of about 2 μg. The sensitivity of immunostaining with peroxidase is 20,000-fold greater, with a limit for detection of about 0.1 ng (Towbin et al., 1979). Although you will not be approaching this limit with your samples, the mass of protein to be loaded in each gel lane should be less than that used previously. Refer to Table 1.4 for guidance.

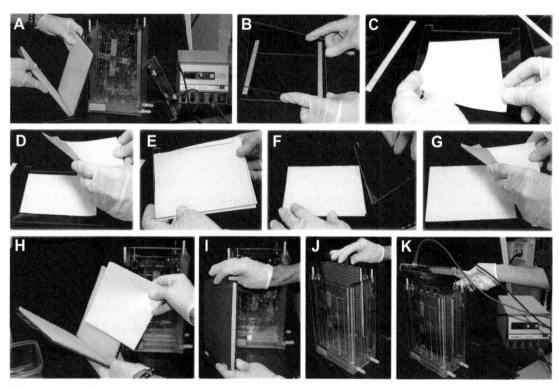

FIGURE 1.29   **Assembly of an apparatus for electrotransfer.**   (A) Typical equipment used for electrophoretic transfer of polypeptides from an acrylamide gel to a sheet of nitrocellulose. From left to right: the *cassette* with porous, spongy pads; chamber for the electrophoretic transfer; lid for the chamber with electrical cables; and electrical power supply. The cassette and chamber shown are model EBU-102 from CBS Scientific (Del Mar, CA). Other designs are available from a variety of companies but this example illustrates the parts of the apparatus common to all such devices. (B) After electrophoretic separation of the polypeptides saturated with dodecyl sulfate, the plates containing the gel of polyacrylamide are removed from the electrophoretic apparatus. They are laid on the bench with the notched plate on the counter, the unnotched plate is pried off, and the plastic spacers are removed from the sides. (C) A pre-wetted sheet of nitrocellulose, the same dimensions as the separating gel, is placed on the separating gel. The sheet of nitrocellulose does not need to cover the stacking gel, only the separating gel, because all of the proteins will have moved into the separating gel. The wells in the stacking gel are used to mark the lanes on the gel onto the sheet of nitrocellulose with a #2 pencil. Note the gloved hands in the picture. (D) Two sheets of pre-wetted Whatman filter paper of the same dimensions as the sheet of nitrocellulose are placed on top of the nitrocellulose. (E) The assembly is inverted on the table top so that the notched plate is facing up. (F) The sandwich of the gel, the sheet of nitrocellulose, and the two sheets of Whatman filter paper is peeled off the plate. (G) Two sheets of pre-wetted Whatman filter paper are placed on the exposed surface of the gel. The gel of polyacrylamide and the sheet of nitrocellulose are now sandwiched between two pairs of Whatman filter sheets. You should note, however, that the two sheets of Whatman filter paper in direct contact with the gel of polyacrylamide have not been in contact with the counter top where they could pick up stray proteins. (H) The sandwich is placed in the cassette. In the device shown, the cassette panels are color coded for proper placement in the chamber: red towards the anode and black towards the cathode. The complexes of polypeptides and dodecyl sulfate are anionic, and they are being transferred from the polyacrylamide to the nitrocellulose, so the nitrocellulose should be next to the anode and the polyacrylamide next to the cathode. If they are reversed, the polypeptides will be transferred away from the nitrocellulose and lost. (I) The cassette is closed. (J) The cassette is placed in the chamber that has been already filled with the buffered solution for the transfer. (K) The electrical cables extending from the chamber lid are plugged into the appropriate ports on the power supply, and the lid is secured on the chamber in the appropriate orientation as indicated by the color coding on the cassette. You should convince yourself that you have chosen the poles properly relative to the gel of polyacrylamide and the sheet of nitrocellulose before you turn on the current. Electrical current is applied, and the complexes between the polypeptides and anions of dodecyl sulfate are transferred to the sheet of nitrocellulose.

TABLE 1.4   Samples of Protein to be Assayed by Electrophoresis Followed by Immunoblotting[a]

	Concentration of Protein (mg mL⁻¹)	Volume (µL)	Mass (µg)
Prestained markers of different lengths[b]	unknown	10	unknown
Concentrated pool from DEAE			5
			10
Concentrated FNR purified by adsorption			0.02
			0.10

[a] To complete this table, use the concentrations of protein from the Bradford assays of your pools.
[b] Volume shown is appropriate if the ColorBurst markers from Sigma-Aldrich are being used. The company does not provide a concentration of protein. ColorBurst markers are already saturated with dodecyl sulfate so they do not need to be treated with dodecyl sulfate before they are loaded.

Procedure (Two Laboratory Sessions)
PART 1: ELECTROPHORESIS AND TRANSFER OF DENATURED PROTEINS TO NITROCELLULOSE

Electrophoresis.

Refer to the exercise in which FNR was submitted to electrophoresis on gels of polyacrylamide cast in solutions of dodecyl sulfate for the preparation of the polyacrylamide gel (12.5% separating gel). The table in that exercise also specifies the amounts of dodecyl sulfate, 2-sulfanylethanol, glycerol, and tracking dye to be added to each sample of protein, as well as the heating of the sample in the bath of boiling water prior to the electrophoresis.

Perform electrophoresis just as you did before. The difference in the present situation is that you will be able to watch the markers separate from each other. Remember to measure the distance the marker dye moved over the period of the electrophoresis. You should measure its movement relative to one or two of the prestained standards, which will be transferred and visible on the sheet of nitrocellulose.

The times required for each step of the electrophoresis on gels of polyacrylamide procedure:

Preparation of the mold for the polyacrylamide (*10 min*).
Preparation of the 12.5% acrylamide separating gel and preparation of the samples for electrophoresis (*40 min*).
Preparation of the 4.5% acrylamide stacking gel (*25 min*).
Loading and performing the electrophoresis (*1.8 h*).
Removal of the gel from the mold (*10 min*).

While the gel is running:

An appropriate volume of buffered solution used for the electrotransfer should be degassed for at least 1 h before being poured into the chamber. This will minimize foaming when the current is applied.

Set up the apparatus for the electrotransfer (20 min).

When the dye front has moved almost to the bottom of the polyacrylamide gel, stop the electrophoresis, and follow the instructions provided below (Figure 1.29) for setting up the electrotransfer. Always wear gloves when handling nitrocellulose because your fingers are covered with protein that would adsorb to it. Remember that your skin itself is almost all protein. Be certain to mark the sheet of nitrocellulose (#2 pencil) for orientation.

Run the electrotransfer overnight (16–22 h) at 100 mA, constant current (usually around 30 V). The reason that constant current is chosen in this instance is to avoid the heating that would occur if the current were too great. During that period of electrotransfer, the polypeptides in the polyacrylamide will migrate laterally out of the polyacrylamide and into the sheet of nitrocellulose.

PART 2: IMMUNODETECTION OF FNR

Perform all steps at room temperature. To avoid using more than is absolutely necessary of the expensive reagents, use only the minimum volume of solution needed to submerge the sheet of nitrocellulose (usually between 10 and 20 mL). To disperse the antiserum and the staining solutions uniformly over the sheet of nitrocellulose, the container holding the sheet should be gently agitated (rotating platform or manually) during each soak.

Peroxidase is used to generate the color at the site of the polypeptide recognized by the antibodies in the antiserum. Azide anion inhibits peroxidase activity by binding tightly to the prosthetic iron ion in its active site, so azide anion must not be present in the solution during the staining steps. Many antisera or solutions of antibodies provided commercially contain 0.01% sodium azide as a preservative, but the dilutions specified below and the washes with phosphate-buffered saline with Tween-20 after the soak with the secondary antibody should lower the concentration of sodium azide to a level at which it will not interfere with the peroxidase.

Removing the sheet of nitrocellulose, with the transferred polypeptides bound to them, from the apparatus for electrotransfer (10 min).

- After electrotransfer, take the cassette out of the chamber, and remove the sandwich. Peel off the Whatman paper from both sides of the sandwich. If the transfer worked, the prestained markers should be bound to the sheet of nitrocellulose; if the sandwich was put into the apparatus backwards, they will not be there. Peel the sheet of nitrocellulose off the gel and begin the procedure for immunostaining.

Processing the sheet of nitrocellulose: blocking, applying antibodies, and washing (2.3 h).

- To block the sheet of nitrocellulose, immerse it in phosphate-buffered saline with Tween-20 and let it soak for 20 min.

- To apply the primary antiserum, prepare a 1:3000 dilution in phosphate-buffered saline with Tween-20. Pour off the blocking solution, then submerge the sheet of nitrocellulose in the diluted antiserum, and soak it for 60 min.

- Wash to remove unbound antiserum using three changes of phosphate-buffered saline with Tween-20 over 20 min.

- To apply the secondary antibody, prepare a 1:400,000 dilution in phosphate-buffered saline with Tween-20, and soak the sheet of nitrocellulose in this solution for 30 min. Note how little of the secondary antibody is needed.

- Wash to remove unbound antiserum using three changes of phosphate-buffered saline with Tween-20 over 10 min.

Staining, recording the image, and storing the sheet of nitrocellulose (1.3 h).

- Prepare a 1:2000 dilution of Extravidin-Peroxidase in phosphate-buffered saline with Tween-20. Submerge the sheet of nitrocellulose in this solution, and soak for 30 min.

- Wash with three changes of phosphate-buffered saline that does not contain Tween-20 over 10 min.

- Submerge the sheet of nitrocellulose in the solution of 3,3′,5,5′-tetramethylbenzidine (wear gloves when handling this solution).

Stained bands (blue) should appear in 5–15 min. When the color of the bands no longer increases, rinse the sheet of nitrocellulose in distilled water for 1 min.

Record a photographic image of the sheet of nitrocellulose while it is still wet. This image of the stained sheet of nitrocellulose is recorded with reflected light. Place the light source above the sheet of nitrocellulose rather than below as you did for the stained gel of polyacrylamide, which, unlike the nitrocellulose, was transparent.

Store the sheet of nitrocellulose in the dark. It will dry out, and the stained bands should be stable for about a week, but because the color fades, it is best to record an image of the stained sheet immediately after the staining.

Analysis of the Data and Questions to Address in Your Report

1. The ColorBurst markers are the polypeptides of myosin (violet), serum albumin (pink), glutamate dehydrogenase (blue), alcohol dehydrogenase (pink), carbonate dehydratase (orange), trypsin inhibitor (blue), lysozyme (pink), and aprotinin (blue). The actual lengths of these polypeptides are 1938, 583, 501, 374, 260, 181, 147, and 58 aa, respectively. The covalent attachment of the dye that gives them their bright colors, however, changes their electrophoretic mobilities, probably by changing the amount of dodecyl sulfate that they bind. Consequently these markers have apparent lengths that depend on the concentration of polyacrylamide in the gel. The apparent lengths reported for a 10–20% acrylamide gradient gel in Table 1.5 seem to be the most appropriate for this exercise. The variation from the actual lengths seems to be at random.

TABLE 1.5   Apparent Lengths of the ColorBurst Electrophoretic Markers[a]

Color	Apparent Length (aa)
Violet	1900
Pink	820
Blue	590
Pink	360
Orange	270
Blue	180
Pink	120
Blue	80

[a] Sigma-Aldrich item C1992. The apparent lengths listed in the table are those for a 10–20% gradient of polyacrylamide.

Create a semilog plot. Plot the negatives of the natural logarithms of the relative mobilities of the markers as a function of their apparent lengths. Make the same plot for their actual lengths.

Hint: The relative mobilities will all be between 0 and 1. Thus, you will be plotting the negatives of natural logarithms of numbers between 0 and 1. In this case, a semilog plot generated by spreadsheet software with a logarithmic axis may be difficult to interpret. Thus it is best to calculate the negatives of the natural logarithms with a calculator and then plot them on a linear scale. You can also use a spreadsheet to calculate the negatives of the natural logarithms and then have a graphing program plot the values.

2. Interpolate from the semilog plot to estimate the length of the polypeptide detected by the anti-FNR antibody. Is it possible to detect a protein multimer with this technique? Is the length you estimated the value expected for the length of the polypeptide of FNR from *S. oleracea*? Does the length estimated agree with the length that you estimated for the major band stained by Coomassie Brilliant Blue?

3. The most critical property of an antiserum is specificity. Comment on the specificity of the primary antiserum used in this experiment. Binding of an antibody to a polypeptide other than one in the antigen used to elicit the immune response in the host animal is *cross-reaction*. There are two types of cross-reaction: nonspecific and specific. If an antibody binds to a polypeptide unrelated in sequence to the expected target, nonspecific cross-reaction is occurring. This can be a serious limitation of immunostaining. Specific cross-reaction occurs when an antibody binds to a polypeptide similar enough in sequence to the expected target to have within its complete sequence of amino acids a sequence or sequences identical or very close to the sequence or sequences of one or more epitopes in the sequence of the target. Did either type of cross-reaction occur in this experiment? Explain why or why not.

4. The samples of protein examined on this immunoblot were also assayed by only electrophoresis of the complexes of their polypeptides with dodecyl sulfate. Discuss whether or not the comparison of both results makes it easier to estimate the effectiveness of the purification by adsorption and elution than evaluation of the polyacrylamide gel stained with Coomassie Brilliant Blue alone?

5. In the technique of *immunohistochemistry* an antigen, usually a particular protein, is detected in a frozen section of a particular tissue by immunostaining. Detergents are not typically used during preparation of the tissue. Suggest an explanation for the fact that some antibodies work well for immunoblots of polypeptides unfolded by dodecyl sulfate but not for immunohistochemistry, while the reverse is true for other antibodies.

6. Both the ELISA, mentioned above in Technical Background, and the immunoblotting you just performed employ the chromogenic product of an enzymatic reaction for detection of the antigen. As explained in the two references provided above, in the ELISA the colored product is soluble and quantified by its absorbance. ELISA is a quantitative method. Discuss whether or not the immunoblot you just made is a quantitative method.

Purification and Characterization of a Recombinant FMN Reductase from *P. leiognathi*

BACKGROUND: FLAVIN REDUCTASES, BIOLUMINESCENCE AND THE *LUX* OPERON

A variety of organisms—fireflies, marine invertebrates, and bacteria—have enzymes in the cytoplasm of their cells or in the cytoplasm of some of their cells that are able to produce light. Most of the luminescent bacteria exist in symbiosis with fish and invertebrates, proliferating within certain specialized organs and making the animals appear to be luminescent. Both the structures of the bioluminescent enzymes and the substrates used by these enzymes vary dramatically among the different phyla, probably because they are unrelated enzymes that evolved independently to catalyze reactions that just happened to produce light. In the related luminescent marine bacteria *Photobacterium leiognathi* and *Vibrio fischeri*, the enzyme that is responsible for the generation of the light is alkanal monooxygenase (FMN-linked) (EC 1.14.14.3).

$$FMNH^- + RCHO + O_2 + H_2O \rightarrow FMN + RCOO^- + 2H_2O + hv \tag{2.1}$$

One of the substrates in the reaction catalyzed by this enzyme is the reduced form of flavin mononucleotide, $FMNH^-$ (Hosseinkhani et al., 2005; Song et al., 2007b). Since the enzymatic reaction generating the light involves the oxidation of $FMNH^-$ to FMN, sustained bioluminescence requires that the FMN produced by the enzyme be reduced continuously back to $FMNH^-$ (Figure i.2B) by FMN reductase (NADH) (Ingelman et al., 1999).

$$FMN + NADH \rightarrow FMNH^- + NAD^+ \tag{2.2}$$

In *V. fischeri* and *P. leiognathi* the alkanal monooxygenase (FMN-linked) is an $\alpha\beta$ heterodimer. The genes, *luxA* and *luxB*, for its two subunits are encoded within the *lux* operon in each of these organisms (Winfrey et al., 1997). This operon also includes the *luxG* gene that encodes FMN reductase (NADH) (Figure 2.1). Thus, the transcription of the *lux* operon and translation of the resulting mRNAs generates both alkanal monooxygenase (FMN-linked) and FMN reductase (NADH) (LuxG) at the same time.

The redox activity of LuxG may be assayed using FMN and NADH as reactants. The coenzyme NADH absorbs maximally at 339 nm but NAD^+ absorbs almost no light at this wavelength. Even though both FMN and $FMNH^-$ absorb light at 339 nm (Figure i.2B), the isosbestic point in their absorption spectra is at 336 nm at pH 8 (Song et al., 2007a). Consequently, as FMN is converted to $FMNH^-$, the absorption from these two forms of flavin at 336 nm does not change. As the oxidation of NADH and the reduction of FMN proceed, the absorbance of the solution at 336 nm decreases. This decrease in A_{336} monitored with a spectrophotometer provides quantification

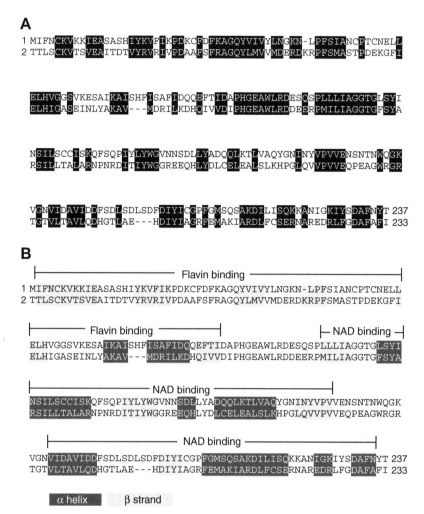

A

```
1 MIFNCKVKKIEASASHIYKVFIKPDKCFDFKAGQYVIVYLNGKN-LPFSIANCPTCNELL
2 TTLSCKVTSVEAITDTVYRVRIVPDAAFSFRAGQYLMVVMDERDKRPFSMASTPDEKGFI

ELHVGGSVKESAIKAISHFISAFIDQQEFTIDAPHGEAWLRDESQSPLLLIAGGTGLSYI
ELHIGASEINLYAKAV---MDRILKDHQIVVDIPHGEAWLRDDEERPMILIAGGTGFSYA

NSILSCCISKQFSQPIYLYWGVNNSDLLYADQQLKTLVAQYGNINYVPVVENSNTNWQGK
RSILLTALARNPNRDITIYWGGREEQHLYDLCELEALSLKHPGLQVVPVVEQPEAGWRGR

VGNVIDAVIDDFSDLSDLSDFDIYICGPFGMSQSAKDILISQKKANIGKIYSDAFNYT 237
TGTVLTAVLQDHGTLAE---HDIYIAGRFEMAKIARDLFCSERNAREDRLFGDAFAFI 233
```

B

```
                    |――――――――――― Flavin binding ――――――――――|
1 MIFNCKVKKIEASASHIYKVFIKPDKCFDFKAGQYVIVYLNGKN-LPFSIANCPTCNELL
2 TTLSCKVTSVEAITDTVYRVRIVPDAAFSFRAGQYLMVVMDERDKRPFSMASTPDEKGFI

  |――――― Flavin binding ―――――|              |― NAD binding ―|
ELHVGGSVKESAIKAISHFISAFIDQQEFTIDAPHGEAWLRDESQSPLLLIAGGTGLSYI
ELHIGASEINLYAKAV---MDRILKDHQIVVDIPHGEAWLRDDEERPMILIAGGTGFSYA

  |――――――――――― NAD binding ―――――――――――|
NSILSCCISKQFSQPIYLYWGVNNSDLLYADQQLKTLVAQYGNINYVPVVENSNTNWQGK
RSILLTALARNPNRDITIYWGGREEQHLYDLCELEALSLKHPGLQVVPVVEQPEAGWRGR

  |――――――――― NAD binding ―――――――――|
VGNVIDAVIDDFSDLSDLSDFDIYICGPFGMSQSAKDILISQKKANIGKIYSDAFNYT 237
TGTVLTAVLQDHGTLAE---HDIYIAGRFEMAKIARDLFCSERNAREDRLFGDAFAFI 233
```

| α helix | β strand |

FIGURE 2.1   **Predicted secondary structure of LuxG from *P. leiognathi*, predicted by comparison to a riboflavin reductase from *Escherichia coli*.**   (A) A comparison of the sequences of amino acids for LuxG FMN reductase (NADH) (top sequence) and Fre riboflavin reductase (NAD(P) H) shows a strong similarity (35.6% identity, 0.9 gap percent). Shading shows identical positions and conservative substitutions. (B) A crystallographic molecular model of Fre from *E. coli* has been determined by X-ray diffraction (Ingelman et al., 1999). The similarity in the sequences of amino acids between LuxG and Fre and their similarity in function have been used to predict the structural domains for LuxG shown here (Nijvipakul et al., 2008).

of the enzymatic activity. Another means of quantifying the rate of the enzymatic reaction is to monitor the decrease of A_{446} as FMN is reduced to FMNH⁻.

A plasmid, pGhis, constructed to express LuxG with a sequence of six histidines at its carboxy terminus (Figure 2.3) has been created by Dr P. Chaiyen's research group (Mahidol University, Thailand). When this plasmid is inserted into *Escherichia coli*, the bacteria express LuxG with this hexahistidine tag at high levels. The recombinant protein that is expressed can be purified in one step by adsorption and elution from a solid phase to which nickel ions are tightly chelated. Imidazoles are strong ligands for Ni^{2+}, so the imidazoles of the six histidines at the carboxy terminus of the enzyme associate tightly as ligands to the free sites on the Ni^{2+} chelated on the solid phase, and the enzyme to which they are covalently attached is, in turn, tightly bound by the adsorbent.

Experiments with this recombinant LuxG, performed in Dr Chaiyen's laboratory, have shown that its characteristics are almost identical to that of the natural enzyme expressed in *P. leiognathi*

(Nijvipakul et al., 2008). Like the natural enzyme in *P. leiognathi*, the recombinant LuxG functions as a homodimer. LuxG is not a flavoenzyme with a prosthetic flavin as a permanent feature of its active site, so the formal hydride ion is transferred between the substrates NAD^+ and FMN directly. Unlike ferredoxin-$NADP^+$ reductase (FNR), which does have a prosthetic flavin that never leaves its active site, the $FMNH^-$ produced in the active site of LuxG has to leave the active site on every turnover of the enzyme because it is a product of the reaction. It then diffuses through the cytoplasm to act as a substrate for alkanal monooxygenase (FMN-linked).

LUXG EXERCISE 1

Expression of LuxG-hexahistidine in *E. coli* and its Release from the Cells by Lysis

Technical Background

Purification of a native enzyme from its natural source is always necessary when questions such as its posttranslational modification, intracellular localization, and the effects of external stimuli on its level of expression are to be addressed. If, however, it is known that the enzyme undergoes no posttranslational modifications, its expression from a recombinant cloned gene or from a recombinant cloned segment of DNA complimentary to its mRNA in a heterologous organism such as *E. coli* is an uncomplicated way to produce large quantities of the enzyme. If the enzyme of interest is difficult or logistically impossible to purify from its natural source in an amount sufficient for the experiments planned, the recombinant method may be the only option. It is often an easier and more efficient way to produce large quantities of the enzyme in any case, but it requires that the mRNA or, if the gene has no introns, the gene encoding the protein, has been isolated.

One problem that can arise in the use of heterologous expression systems for a recombinant version of the protein is toxicity. A foreign protein may slow or stop the cell cycle of the host organism (Studier, 2005; Studier et al., 1990). The inducible expression system described below was created to address this issue. Expression of the recombinant protein is repressed as the bacterial culture is grown into the period of exponential growth. In many cases, induction of expression for a few hours during exponential growth is not harmful to the host cells and large amounts of the recombinant protein can be generated.

Expression of recombinant LuxG from the pGhis plasmid (Nijvipakul et al., 2008) is regulated by the *lac* repressor, and the *luxG* gene is transcribed by DNA-directed RNA polymerase from bacteriophage T7 (T7 RNA polymerase). Regulation of the *lac* operon in wild type *E. coli* is discussed in textbooks (Dale et al., 2011; Krebs et al., 2013; Watson, 2008). The regulation of the *lac* operon by the *lac* repressor is an important mechanism for the regulation of the expression of the *lac* operon when the bacteria are living in their accustomed environment. In the case of the expression of recombinant proteins, in which it has artificial function, the *lac* operator, which is a short segment of DNA to which the *lac* repressor is bound when it is repressing the transcription of the *lac* operon, is present in the plasmid just upstream of the *luxG* gene, and as long as *lac* repressor is bound to it, the transcription of the *luxG* gene is repressed. The gene encoding T7 RNA polymerase has been incorporated into the DNA of the strain of *E. coli* used to express the protein. The strain of *E. coli* used to express LuxG is also designed so that the gene for T7 RNA polymerase is only transcribed when the *lac* repressor is inactivated, so that both the *luxG* gene and the gene for T7 RNA polymerase are turned on simultaneously. DNA-Directed RNA polymerase from bacteriophage T7 transcribes DNA much more rapidly than the normal DNA-directed RNA polymerase in the cells of *E. coli*, and this rapid rate causes its transcription of the *luxG* gene to monopolize transcription in the cell once its gene and the *luxG* gene are derepressed simultaneously.

This strategy was developed by academic investigators (Studier, 2005; Studier et al., 1990) and the Novagen company then created a series of pET (plasmid for expression by T7 RNA polymerase) vectors (Novagen, 1998) and host strains (Novagen, 2006b) for regulating and accomplishing expression using this strategy (Figures 2.2 and 2.3). Novagen also markets reagents for a method of lysing of bacteria that minimizes the inactivation of the protein being expressed (Novagen, 2007).

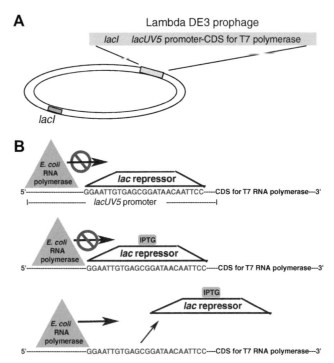

FIGURE 2.2   **Regulated expression of T7 RNA polymerase inserted into the genome of the Tuner(DE3) strain of *E. coli*.**   The genotype of Tuner(DE3) is F⁻*ompT hsdS*ᴮ(rᴮ⁻ mᴮ⁻) *gal dcm lacY1*(DE3). Tuner(DE3) is the strain of *E. coli* used to express LuxG. The *lacI* gene in wild type *E. coli*, which is retained in Tuner(DE3), encodes the *lac* repressor. Tuner(DE3) is a strain of *E. coli* that is a *lysogen* of the lambda DE3 prophage. In a lysogen such as Tuner(DE3), a modified form of the DNA of the lambda bacteriophage, in this case the lambda DE3 prophage, is stably integrated into its chromosome. Within the modified lambda DNA incorporated into the chromosome of Tuner(DE3) is included an additional copy of the *lacI* gene and the coding sequence (CDS) for T7 RNA polymerase. In the integrated prophage, a mutant version of the *lac* promoter containing the *lac* operator from the *lacUV5* strain of *E. coli* is fused upstream of the coding sequence for T7 RNA polymerase. (A) A schematic drawing showing the endogenous *lacI* gene on the *E. coli* chromosome, and the DE3 prophage just described inserted into the chromosome. (B) Detail of the control region upstream of the gene for T7 RNA polymerase. At the 3′ end of the *lacUV5* promoter is the *lac* operator, the sequence of nucleotides recognized by the *lac* repressor, to which the *lac* repressor binds tightly preventing any T7 RNA polymerase from being expressed. The DNA-directed RNA polymerase from the host *E. coli* is bound to the promoter but its procession is blocked by the repressor. Isopropylthiogalactoside (IPTG) binds to the repressor, changing its conformation and affinity for the operator and causing the repressor to dissociate from the template. After it dissociates, transcription of T7 RNA polymerase can proceed.

The optimal temperature for the growth of *E. coli* is 37 °C, and this temperature is used for the expression of many recombinant proteins. A common problem with expression of recombinant proteins in bacteria is the formation of insoluble and useless aggregates of the protein that is being expressed. One cause of this aggregation is that systems for expression are designed to produce the protein of interest at high levels; levels higher than most of the endogenous proteins in the bacterium. Undesirable aggregation was encountered in the initial use of pGhis to express LuxG-hexahistidine (Nijvipakul et al., 2008). A modified procedure, however, in which the bacteria were cultured at a lower temperature (16 °C) minimized the problem. For this reason, the procedure you will be using stipulates 16 °C as the temperature for the culture during the induction with isopropylthiogalactoside.

In order to purify the recombinant protein, the bacteria that are serving as host for the expression must first be lysed. All living cells are enclosed completely by a fragile *plasma membrane* consisting of a lipid bilayer of phospholipids and glycolipids. There is great variation, however, among different cell types and tissues in the structures that then enclose and protect this delicate plasma membrane (Lehninger et al., 2008; Ninfa et al., 2010; Voet and Voet, 2010). In mammalian tissues, a major portion or all of the surface of each cell is bordered by other similar cells, so that each cell is supported and

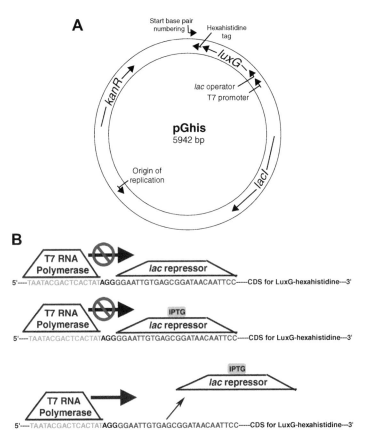

FIGURE 2.3   **Regulated expression of recombinant LuxG with the hexahistidine tag in the Tuner(DE3) strain of *E. coli* bacteria.**   The synthesis and initial use of the pGhis plasmid has been previously reported (Nijvipakul et al., 2008). (A) A map of the genes in the pGhis plasmid, based on the sequence of the pET24b vector and the protein-encoding sequence of *luxG* that are available in public databases. A plasmid is a small circular piece of DNA that can be replicated on its own from an origin of replication. The positions of the relevant genes in the plasmid and their direction of transcription are noted in the diagram. In addition to the coding sequence (CDS) for LuxG-hexahistidine, the pGhis plasmid contains a copy of the *lacI* gene encoding the *lac* repressor, and a gene conferring kanamycin resistance on the host so that only bacteria containing the plasmid will grow in the presence of kanamycin. (B) Events that occur at the promoter for the *luxG* gene on the plasmid. Only T7 RNA polymerase, and not DNA-directed RNA polymerase from *E. coli*, binds the promoter for the transcription of the *luxG* gene. Before induction, however, the repressor blocks procession of the polymerase along the template. Isopropylthiogalactoside (IPTG) binds to the repressor, changing its conformation and affinity for the operator and causing the repressor to dissociate from the template. After it dissociates, transcription of the *luxG* gene can proceed at the high rate achieved by T7 RNA polymerase.

protected by its neighbors. Mammalian cells are also protected from the environment by the *extracellular matrix*, a network of proteins, glycoproteins, and polysaccharides. *Cell walls* surround the cells in the tissues of plants. A cell wall is a layer of cellulose, which is a polymer of glucose with (β1,4) glycosidic linkages. In contrast to these extracellular layers of protection in multicellular eukaryotes, each bacterium is surrounded by a network of *peptidoglycan*. Because bacteria are single cells that are fully exposed to the environment, this layer of peptidoglycan is tougher than the layers surrounding eukaryotic cells. Thus, lysis of bacteria is usually more difficult than lysis of eukaryotic cells.

The protocol you will be using includes the treatment of the bacteria with the enzyme lysozyme to partially digest the cell wall, followed by sonication, to disrupt the cells of *E. coli* completely. The sonifier produces high-frequency waves of sound that create shear forces strong enough to lyse the bacteria and fragment their plasma membranes and cell walls.

Experimental Plan

Expression of recombinant protein LuxG-hexahistidine from the pGhis plasmid will be induced by addition of isopropylthiogalactoside to a culture of Tuner(DE3) bacterial cells that carry the plasmid pGhis (Figures 2.2 and 2.3). The recombinant LuxG-hexahistidine will be released from the cells by a combination of enzymatic digestion of the cell walls and sonication. The concentration of the proteins that have been released will be quantified with the Bradford assay.

Reagents, Supplies, and Equipment

The pGhis plasmid for expression of LuxG-hexahistidine. This plasmid has the coding sequence (717 bp) of *luxG* from *P. leiognathi* with a coding sequence for a hexahistidine tag at its carboxy terminus (Figure 2.2) inserted into the plasmid pET24b (5309 bp), which includes the *lacI* gene and the *kanR* gene to produce resistance to kanamycin. The final plasmid, with the insert, contains 5942 bp.

The Tuner(DE3) host strain of *E. coli* with the genotype F⁻*ompT hsdS*$_B$(r$_B$⁻ m$_B$⁻) *gal dcm lacY1* (DE3).

Sterile LB broth with the antibiotic kanamycin (30 μg mL^{-1}): 250 mL in a 1 L Erlenmeyer flask.

100 mM (24 mg mL^{-1}) isopropylthiogalactoside (molar mass = 238 g mol^{-1}) in water, sterilized by filtration, to induce expression.

60 mM (10 mg mL^{-1}) phenylmethanesulfonyl fluoride (molar mass = 174 g mol^{-1}) in ethanol, stored at −20 °C, to inhibit serine endopeptidases.

Resuspension solution: 10% glycerol, 100 μM phenylmethanesulfonyl fluoride, 50 mM Tris chloride, pH 8.

1000 units μL^{-1} rlysozyme™ (Novagen), a lysozyme of undisclosed origin.

Reagent for the Bradford assay: 0.01% (w/v) Coomassie Brilliant Blue G-250, 4.75% (v/v) ethanol, 8.5% (v/v) phosphoric acid.

Standard for Bradford assay: bovine immunoglobulin G, 0.1 or 1 mg mL^{-1}.

Screw-cap tubes and bottles for centrifugation (Figure 1.6):

Centrifuge bottle (250 mL) with rubber O-ring in cap (to ensure tight seal) for the low-speed spins (≤6000 rpm) to pellet cells.

Clear plastic conical tubes (15 mL; disposable) with thin (1 mm) walls and a conical bottom for the low-speed (2000 rpm) spins and sonication.

Opaque plastic Oak Ridge tubes (38 mL; reusable) with thick (2–3 mm) walls and a round bottom for the high-speed (13,000 rpm) spins.

Materials for spectrophotometry and labeling of test tubes:

Disposable plastic cuvettes (0.5 or 1 mL) for the bacterial cultures.

Multiwell plate (for plate reader, Figure 1.13 and Appendix I), 0.5 mL disposable plastic cuvettes (for conventional spectrometer), or glass culture tubes (for Spec 20) for use in Bradford assay.

Markers (VWR or Fisher) for labeling tubes to be quick-frozen. The markers must be alcohol-proof and waterproof with black ink (blue or other colors will wash off).

Equipment:

Shaking incubators at 16 and 37 °C.

Beckman J2 refrigerated preparative centrifuge and JA-14 rotor (or equivalent).

Sonifier (Figure 2.4, Branson 450 or equivalent).

Spectrophotometers: automated plate reader (Figure 1.13), conventional spectrophotometer, or Spectronic 20D+ (Spec 20).

Precautions

You should be supervised by the instructor or teaching assistant while using the preparative centrifuge. Be certain the sample bottles are balanced and the rotor is secure before the spin is initiated.

The high frequency sound of the sonifier can damage your ears. Wear ear protection.

The Bradford reagent contains phosphoric acid that can erode flesh and Coomassie Brilliant Blue G-250 that is toxic. Wear gloves.

Procedure (One Extended Laboratory Session and One Regular Session)

PART 1: EXPRESS LUXG-HEXAHISTIDINE, AND PELLET AND FREEZE CELLS

This part cannot be completed in a typical laboratory session. The most practical approach is for the instructor or teaching assistant to inoculate the cultures to be grown at 37 °C in the morning. The students, after they arrive in the laboratory, should monitor the cell densities in the early afternoon and at the appropriate time transfer the cells to 16 °C and induce the expression. The students should then come back to the laboratory in the evening to harvest and freeze the cells.

Monitor growth of cells and induce with isopropylthiogalactoside (2 h of student work in early afternoon).

At 37 °C, the time it takes for a culture of Tuner(DE3) bacteria/pGhis to double in cellular density is 30 min but at 16 °C it takes 7 h. A culture with a cellular density of 1×10^8 cells mL^{-1} has an A_{600} of 0.6 ($\varepsilon_{600} = 6 \times 10^{-9}$ mL cell^{-1} cm^{-1}). In this case, you are not actually measuring the absorption of light; you are measuring the diminution in the light passing straight through the sample to the detector of the spectrophotometer caused by the scattering of the light by the cloudy suspension of bacterial cells, much as clouds scatter sunlight and dim the disc of the sun.

The culture will be grown at 37 °C until it reaches its induction density. At the induction density, the A_{600} should be between 0.4 and 0.5. When the cells reach this density, the inducer, isopropylthiogalactoside, will be added. The incubation will then be shifted to 16 °C and allowed to proceed for 7 h. The A_{600} should double during this incubation.

An ideal negative control is a culture of the Tuner (DE3)/pGhis bacteria incubated at the same two temperatures for the same periods of time but not induced with isopropylthiogalactoside. The instructor will prepare and lyse such a culture to provide control samples for the electrophoretic analysis and the measurement of enzymatic activity.

- Four hours before the session is to begin in earnest, cultures for each group of students each containing 250 mL of LB broth with the antibiotic, kanamycin, will be inoculated with three colonies of Tuner(DE3)/pGhis growing on an agar plate, each colony about 1 mm in diameter, and the cultures will then be incubated at 37 °C with shaking.

- The culture should reach the density necessary for induction in about 5 h. Begin assaying absorbance after 4 h of incubation. It is best to use disposable plastic cuvettes (0.5–1 mL) in a conventional spectrophotometer to minimize the amount of culture sacrificed. For this purpose, the disposable cuvettes can be used more than once since precision is not required.

- When density suggested for induction is attained, cool the culture to 16 °C, and then add isopropylthiogalactoside **aseptically** to a final concentration of 0.4 mM.

- Grow at 16 °C for 7 h to achieve full expression.

- Assay A_{600} again to confirm that the bacteria continued to grow during the induction.

Spin down the cells and resuspend them (1 h of student work in the evening).

- Transfer the culture to a centrifuge bottle, and balance it with another similar bottle by transferring culture between the two bottles. Be certain the cap of the centrifuge bottle is tight enough to prevent leakage during the centrifugation. You definitely want to avoid contaminating the rotor with bacteria because you will be required to clean a rotor contaminated with bacteria if there is any leakage!

- Spin down the cells at 6000 rpm in a JA-14 rotor ($5500 \times g$) for 20 min at 4 °C.

- Resuspend the pellet of cells in 10 mL of 10% glycerol, 100 μM phenylmethanesulfonyl fluoride, 50 mM Tris chloride, pH 8. The concentration of cells after the resuspension should be approximately 5×10^6 cells μL^{-1}. The glycerol in the resuspension prohibits the formation of ice crystals during the freezing. When ice crystals form, they are necessarily pure water, and the other solutes are excluded from the crystals and become highly concentrated in the solution surrounding the crystals. This dramatic increase in the concentration of protein in the excluded solution causes proteins to denature and precipitate. The glycerol causes the solution to freeze as a glass. Glasses, like window glass, are noncrystalline, frozen solutions.

- Transfer the resuspended cells to a 15 mL screw-cap tube with a conical bottom. Do not use a larger tube! A tube with a volume of 15 mL is ideal for sonication.

- It is imperative that you label each of the tubes with your initials, the date, and the type of tissue. There are two types of marking pens generally used in labs. The Sharpies are fine for assay tubes that are only needed for a single experiment, then discarded. For any tube that will be frozen in a slurry of dry ice in isopropanol you must use the alcohol-resistant Fisher or VWR laboratory markers! The isopropanol will dissolve the ink from other types of marker. Black ink sticks are better than blue or other colors. Write directly on the tube; tape will fall off in the freezer.

Freeze the sample in the tube in a slurry of dry ice and isopropanol and store it in the freezer at a temperature of −20 °C until you are ready to sonicate the cells. A freezer with a temperature of −20 °C is sufficient for the storage of all samples of LuxG-hexahistidine between sessions.

PART 2: LYSE THE TUNER(DE3)/PGHIS CELLS AND ASSAY THE CONCENTRATION OF PROTEIN (ONE LABORATORY SESSION)

Enzymatic digestion of cell wall (30 min).

A combination of enzymatic and mechanical disruption will be used to lyse the cells of *E. coli*. After the cells are thawed, rlysozyme is added to the suspension of bacteria to digest the proteoglycan of their cell walls (Novagen, 2007). Further breakdown of the cell wall and disruption of the plasma membrane is then achieved by exposing the preparation to pulses of sound of short wavelength with a sonifier (Figure 2.4). A clearing spin is performed to pellet the fragments of cell wall and membrane (and proteins linked to these structures). The soluble LuxG-hexahistidine remains in the supernate.

Novagen recommends 5000 units of rlysozyme for every gram of cells. The culture that you have made should contain around 2 g of cells, so use 10,000 units.

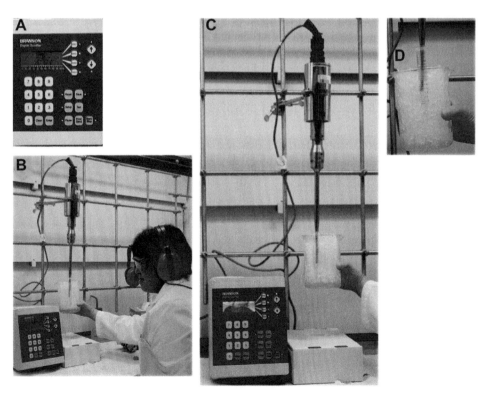

FIGURE 2.4   **A sonifier generates shear forces strong enough to disrupt the plasma membranes and the cell walls of bacteria and lyse the cells.**   (A) The control panel is used to set the program. (B) A fume hood is the ideal location for this device because it provides some protection from splashing and the sound. The operator should wear protection for both eyes and ears. (C) To minimize the unavoidable increase in the temperature of the sample, it is placed in a beaker of ice during the pulses. (D) The 10 mL sample is in a 15 mL plastic conical tube. A larger tube increases the opportunity for undesired splashing and foaming. A *microtip* sonifier probe that narrows to 0.3 cm is used. The tip should be submerged to the 5 mL mark on the tube. Avoid any contact between the probe and the bottom of the tube as this may cause the tube to crack.

- Remove the resuspended bacteria from the freezer, and thaw the frozen suspension at room temperature.

- Add the appropriate amount of rlysozyme and incubate at 30 °C for 15 min.

Sonication (wear protection for your ears) (10 min).

   This step is tricky. If you are not careful, you will not produce enough enzyme to complete the remaining exercises involving LuxG. It is critical that: (1) at least 90% of the bacteria are lysed by the sonication, (2) splashing is minimized so that the cells are consistently exposed to the full power of the sonifier and that loss of sample is avoided, and (3) the enzyme is not denatured by an increase in temperature resulting from the absorption of sonic energy by the solution.

- Do two rounds of 20 pulses with an amplitude of 50 W (50% maximum power) and a duty cycle with pulse on for 0.7 s and pulse off for 0.3 s (Figure 2.4).

- In spite of being on ice during this procedure, the sample will warm up. After each round of sonication, leave the sample on ice for at least 5 min to bring the temperature back down before proceeding.

- After the sonication is complete, transfer the sample to a 38 mL Oak Ridge tube that has been sitting on ice.

Centrifugation of the sonicated lysate to remove membranes and membrane-associated proteins (30 min).

- Spin the sonicated lysate at 13,000 rpm in JA-14 rotor (26,000 × g) for 20 min at 4 °C. Transfer the supernate to a 15 mL clear plastic conical tube. The solution should be pale yellow. If it shows no color, the cells were not lysed. This is the *initial supernate.*

- Examine the pellet for inclusion bodies. Exogenous proteins expressed in *E. coli* sometimes become sequestered in these structures if the tertiary structure has not formed properly because the expression was too vigorous (Novagen, 2006b). The portion of the pellet consisting of membranes and cell walls will be light brown because of the cytochromes in the membranes, whereas inclusion bodies are white. Regardless of how much of the pellet comprises inclusion bodies, the pellet is discarded because any LuxG-hexahistidine precipitated in inclusion bodies will be inactive and insoluble. Even if some of the active soluble enzyme is trapped in the pellet, the majority should be in the supernate.

Remove samples of the initial supernate (15 min).

- A sample (100 μL) to be used to determine the concentration of protein by the Bradford assay.

- A sample (100 μL) to be used for electrophoresis.

- A sample (500 μL) for assaying enzymatic activity.

One microliter of the supernate should contain the cytoplasm from about 5×10^6 cells. The lysate from 15×10^6 cells will usually provide enough protein for a lane on an electrophoretic gel. Although the $100 \, \mu L$ sample reserved for electrophoresis is probably more than necessary, it is best to reserve this volume in case the growth of the culture or the lysis or both of these procedures were not optimal. For the assay of enzymatic activity, about $30 \, \mu L$ of lysate should be enough for each 3 mL tube in a Spec 20.

The remainder of the lysate should be frozen as one sample in a slurry of dry ice in isopropanol and stored at $-20 \, °C$. This will be submitted to purification by affinity adsorption. Recall that the lysate already contains glycerol.

If you will be performing the Bradford assay during this period, keep that sample on ice. Freeze the other two samples in a slurry of dry ice in isopropanol and store them at $-20 \, °C$.

Determine the concentration of protein with the Bradford assay (1 h).

Perform the Bradford assay as you have done before. Make a 1:10 dilution of the initial supernate with water.

If you are using a multiwell plate with a reaction volume of 0.2 mL, try 2 and $5 \, \mu L$ samples of the diluted supernate.

If you are using cuvettes and a conventional spectrophotometer with 0.5 mL cuvettes, try 5 and $10 \, \mu L$ samples of the diluted supernate.

If you are using glass tubes in a Spec 20 with reaction volumes of 3 mL, try 30 and $60 \, \mu L$ samples of the diluted supernate.

If none of the values for A_{595} of these samples are within the linear range of the absorbances of the standard, adjust the volume of the sample added to the assay and try again.

Analysis of the Data and Questions to Address in Your Report

1. Calculate the concentration and total mass of protein in the initial supernate.

2. Does it appear as if the bacterial lysis worked well? What evidence makes you think that it did?

3. A preparation of plasmids is made by lysing cells of *E. coli* harboring many copies of a recombinant plasmid and purifying these small circular molecules of DNA. There are many procedures for purifying plasmids, but some use dodecyl sulfate to lyse the cells. Discuss whether or not it would it be appropriate to use dodecyl sulfate to lyse the cells before purifying an enzyme such as LuxG from *E. coli*.

4. In the expression system used here, T7 RNA polymerase should not be expressed until isopropylthiogalactoside is added to the culture. Regulation of transcription by the *lac* repressor is not always perfect, and it is possible that a low level of this polymerase might be expressed prior to addition of isopropylthiogalactoside. Discuss whether or not this low level of T7 RNA polymerase is likely to generate significant expression of LuxG-hexahistidine and whether or not this would matter.

LuxG EXERCISE 2

Affinity Adsorption of LuxG-hexahistidine to a Solid Phase to which Ni^{2+} is Chelated

Technical Background and Experimental Plan

The imidazolyl groups of the six histidines that have been added to the carboxy terminus of LuxG will bind the Ni^{2+} chelated to the solid phase (Figure 2.5). In this instance, the solid phase has already been poured into a column (Boyer, 2012; Novagen, 2006a). Most, if not all, of the other proteins in the lysate of the bacterial cells flow through the solid phase without associating with it. The solid phase is washed with a buffered solution containing an intermediate concentration of imidazole to release any proteins loosely bound to the Ni^{2+} by single imidazolyl groups on their surface. The tagged LuxG is then eluted with a buffered solution containing a high concentration of imidazole. Although the enzymatic activity of LuxG is stabilized by the presence of a reducing agent, usually dithiothreitol, this is not added until after the enzyme has been eluted from the solid phase because the sulfanyl groups on dithiothreitol have an even higher affinity for nickel than do imidazolyl groups, and they would occupy tightly all of the vacant sites for ligands on the Ni^{2+} and prevent the LuxG-hexahistidine from associating.

Reagents, Supplies, and Equipment

Iminodiacetic acid agarose with a capacity of approximately 20 μmol Ni^{2+} mL^{-1} of bed volume.

50 mM $NiSO_4$.

Three solutions containing imidazole at 20 mM (for the initial adsorption), 50 mM (for washing), and 250 mM (for the elution), respectively. Each of the solutions also contains 10% glycerol, 250 mM NaCl, 50 mM Tris chloride, pH 8.

1 M dithiothreitol.

Reagents and spectrophotometric supplies for the Bradford assay, as in FNR Exercise 2.

A glass column with a diameter of 1 cm and a height of 10 cm (as in FNR Exercise 3, Figure 1.17) with a stopcock. Neither tubing nor a fraction collector is necessary.

Amicon Ultra-4 10K ultrafiltration device with a capacity of 4 mL (Figure 1.16).

Spectrophotometer: automated plate reader, conventional spectrophotometer, or Spec 20 (Figure 1.13).

Precautions

The Bradford reagent contains phosphoric acid that can erode flesh and Coomassie Brilliant Blue G-250 that is toxic. Wear gloves.

Procedure (One Laboratory Session)

Purification by adsorption to chelated Ni^{2+} (2 h).

Usually, about 4% of the protein in the lysate of the Tuner/pGhis bacterial cells is LuxG-hexahistidine. For this amount of protein tagged with hexahistidine, a solid phase with a bed volume of 1 mL should be sufficient.

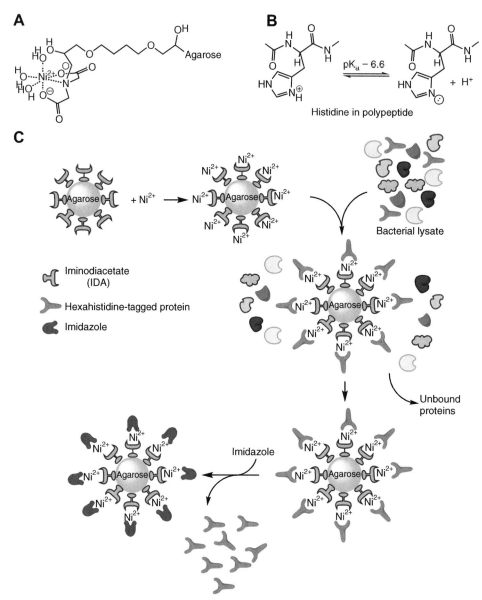

FIGURE 2.5 **Purification of a protein tagged with hexahistidine by adsorption to Ni²⁺ chelated to a solid phase.** (A) Ions of Ni²⁺ chelated by iminodiacetate that, in turn, is covalently attached to agarose. In the drawing, molecules of water occupy the sites on the Ni²⁺ that are the sites to which the imidazolyl groups in the six histidines at the carboxy terminus of LuxG associate. Imidazolyl groups are stronger ligands for Ni²⁺ than are molecules of water, and imidazolyl groups exchange with the waters to form tighter complexes. The hexahistidinyl segment on LuxG-hexahistidine must bind to at least two and probably all three of the vacant sites on a chelated Ni²⁺ because it takes 0.25 M free imidazole to displace it from the solid phase even though the molar concentration of LuxG-hexahistidine is less than 10 μM. (B) The imidazolyl group of histidine is a weak base with a pKa of 6.6. Because the protonated form of an imidazolyl group lacks the lone pair of electrons on the nitrogen that is the ligand to the Ni²⁺, adsorption of LuxG-hexahistidine to the solid phase is performed at pH 8. (C) A protein that is tagged with a hexahistidine segment at either its amino terminus or its carboxy terminus adsorbs to the solid phase because of the affinity of its imidazolyl groups for the Ni²⁺ ions. Proteins that do not have hexahistidinyl segments at one or the other of their termini do not adsorb to the solid phase and are washed away. Imidazole, free in solution, can displace LuxG-hexahistidine from the solid phase because it is present at a high enough concentration (0.25 M) to compete with the hexahistidine tag for the vacant sites on the Ni²⁺ ions (Nijvipakul et al., 2008; Novagen, 2006a). After it has been displaced and is free in solution, the LuxG-hexahistidine is eluted from the solid phase.

Set up the column in a cold room (Figure 1.17). Fill it halfway with water, and then pipette in 1 mL of the solid phase.

After the solid phase settles, estimate the bed volume, and add more if necessary.

Fill the iminodiacetate groups on the solid phase with Ni^{2+} by flowing 5 mL (five bed volumes) of 50 mM $NiSO_4$ through it; the solid phase will turn green as the Ni^{2+} is chelated. After saturating the solid phase with Ni^{2+}, rinse it with 10 mL of 20 mM imidazole, 10% glycerol, 250 mM NaCl, 50 mM Tris chloride, pH 8, to prepare it for the association of the LuxG-hexahistidine.

Apply the initial supernate from the lysate to the column, then follow with 30 mL (30 bed volumes) of 20 mM imidazole, 10% glycerol, 250 mM NaCl, 50 mM Tris chloride, pH 8, to wash most of the untagged protein through.

Wash the solid phase with 5 mL (five bed volumes) of 50 mM imidazole, 10% glycerol, 250 mM NaCl, 50 mM Tris chloride, pH 8, to ensure that all of the untagged protein has been removed. Collect this wash in 1 mL fractions (manually, using snap-cap tubes in a rack) just in case some of the enzyme has also eluted.

Elute the tagged LuxG-hexahistidine with 5 mL (five bed volumes) of 250 mM imidazole, 10% glycerol, 250 mM NaCl, 50 mM Tris chloride, pH 8, collecting 1 mL fractions manually. The tagged protein is usually distributed in three fractions.

Close the stopcock so that solid phase does not dry out in case you need to reuse it. The solid phase should always be stored in its hydrated state.

Assay of the concentration of protein in the fractions from the column (40 min).

Use the Bradford assay, as described in FNR Exercise 2, to measure the concentration of protein in the last three fractions of the wash with 50 mM imidazole and all of the fractions of the material eluted with 250 mM imidazole.

Volumes for the samples from each fraction to be assayed:

If you are using a multiwell plate, use 10 μL samples in 0.2 mL assays.

If you are using cuvettes and a conventional spectrophotometer, use 25 μL samples in 0.5 mL assays.

If you are using glass tubes and a Spec 20, use 60 μL samples in 3 mL assays.

Calculate the total mass of protein in each fraction.

Pool separately the fractions containing the majority of the protein eluted with 50 mM imidazole, if there is any protein in these fractions, and the fractions containing the majority of the protein eluted with 250 mM imidazole. Base the pooling on the raw absorbance of the Bradford assays. Measure the volume of the pools, and add dithiothreitol to a final concentration of 1 mM. The protein is now in a solution containing either 50 mM imidazole or 250 mM imidazole, as well as 250 mM NaCl, 50 mM Tris chloride pH 8, 10% glycerol, and 1 mM dithiothreitol.

• Calculate the concentration of protein and the total amount of protein in these two pools.

Increase by ultrafiltration the concentration of protein in the pool of LuxG-hexahistidine purified by adsorption and elution (30 min).

You will concentrate only the pool of the fractions containing the protein eluted at 250 mM imidazole because this is supposed to be where LuxG-hexahistidine is located. The Ultra-4 device will accept up to 4 mL of solution (Figure 1.16). If the volume of the pool of LuxG-hexahistidine from the purification by affinity adsorption is greater than this, do multiple loadings. Because its effective diameter is greater than the effective diameter of the pores in the filter, all of the LuxG-hexahistidine should be retained in the upper chamber. Because the other molecules in the sample that are not proteins are small enough to pass through the pores, their concentrations are not affected by this procedure.

• Add the pool from the elution with 250 mM imidazole or the two successive portions of this pool to the upper chamber, and spin at 7000 rpm in the JA-14 rotor (7500 × *g*) for 10 min at 4 °C.

• You should reduce the volume of the pool to around 0.3 mL. If the volume of the pool has been decreased sufficiently in the first centrifugation, use a P-200 pipette to transfer it to a test tube. If the volume still needs to be reduced further, do another spin for an appropriate length of time before recovering the sample of protein. Rinse the upper chamber with 50 μL of buffer and add the rinse to the sample of protein.

• Record the final volume. Split the sample into four equal aliquots, and freeze them in a slurry of dry ice in isopropanol, and store them at −20 °C. At the same time, freeze the pool of the protein eluted with 50 mM imidazole, if there was a pool that contained protein, just in case it happens to contain the LuxG-hexahistidine.

Analysis of the Data and Questions to Address in Your Report

1. Prepare a graph of the concentration of protein in the fractions from the column by plotting the concentration of protein in milligrams milliliter^{-1} as a function of the fraction number.

2. Calculate the concentration of protein present in the pool of protein purified by affinity adsorption and elution, and concentrated by ultrafiltration under centrifugal force.

3. If all of the protein in the pool is LuxG-hexahistidine and if all of the LuxG-hexahistidine was recovered, what percentage of the protein in the initial supernate was the enzyme?

4. How could you evaluate whether or not the majority of the recombinant LuxG-hexahistidine was adsorbed and then eluted from the solid phase to which Ni^{2+} was chelated?

5. The solid phase that had been charged with Ni^{2+} is rinsed with enough imidazole to occupy all of the sites on the Ni^{2+} with imidazoles before the lysate is added. What does this say about the affinity of the hexahistidine tag for Ni^{2+}? Why might this be the case?

6. In the exercises concerned with FNR and LDH, the affinity of each of these proteins for a solid phase to which a ligand for the respective enzyme has been covalently attached is exploited

for purification. A recombinant expression system allows the gene or DNA complementary to the mRNA encoding the protein of interest to be fused to a short sequence of DNA encoding a tag that facilitates purification by affinity adsorption. In this exercise, a hexahistidine tag was used to purify LuxG. Alternative tags that have been used include a segment of 11 amino acids from the *gene 10* protein of phage T7, a segment of 8 amino acids that has been selected by trial and error for its ability to bind tightly to the protein streptavidin, and a segment of 15 amino acids from bovine ribonuclease for which antibodies are available (Novagen, 2012).

a. The variety of tags available emphasizes their importance. Explain why it is useful to have these options for the purification of a recombinant protein.

b. Consider the possible effects of such tags. Discuss how the presence of a tag on a recombinant enzyme might affect interpretation of assays for enzymatic activity.

LuxG EXERCISE 3

Electrophoresis of the Polypeptides in the Lysate and Purified LuxG-hexahistidine

Experimental Plan

You will compare by electrophoresis the following samples: (1) the supernate of a lysate of uninduced Tuner(DE3)/pGhis bacteria prepared by the teaching assistant, (2) the supernate of the lysate from induced Tuner(DE3)/pGhis from LuxG Exercise 1, and (3) LuxG-hexahistidine purified by adsorption and elution from the solid phase of chelated Ni^{2+} in LuxG Exercise 2. The polypeptides in the samples will be unfolded and coated with dodecyl sulfate ions as before. The gel of polyacrylamide will be the same as before, with a stacking gel and a separating gel both cast in the same buffered solutions containing dodecyl sulfate. Although the stacking gel is identical to that used for FNR, the separating gel will contain 15% polyacrylamide rather than 12.5% polyacrylamide because the polypeptide of LuxG is shorter than that of FNR.

Samples of protein to be loaded:

A sample of uninduced Tuner(DE3)/pGhis lysate containing 25 μg protein.

A sample of the supernate of the lysate from induced Tuner(DE3)/pGhis containing at least 75 μg of protein.

A sample of LuxG-hexahistidine purified by adsorption and elution containing at least 15 μg protein.

A sample of standard polypeptides containing 16 μg of protein.

Reagents, Supplies, and Equipment

Reagents for unfolding the polypeptides with dodecyl sulfate, for the electrophoresis, for loading the gels, and for staining the gels as well as all of the supplies and equipment needed for the electrophoresis are the same as those described in FNR Exercise 4.

Precautions

As before, the apparatus for the electrophoresis has a safety interlock, but you still need to be careful to avoid any direct contact with electrical current because the ionic solutions make your skin an even better conductor than it usually is.

Simply Blue stain contains Coomassie Brilliant Blue G-250 which is toxic, so handle it with gloves. Although the staining solution does not contain an organic solvent, it is best to place the staining trays overnight in the fume hood.

Acrylamide is a neurotoxin that can be adsorbed through the skin: wear gloves and use pipette bulbs. Never pour acrylamide down the drain. Polymerized acrylamide is much less toxic. Polymerize any unused acrylamide and then toss the solid into the trash.

Procedure (One Laboratory Session)

Refer to the figures in FNR Exercise 4 for guidance in preparing and loading the gel and for the process of electrophoresis. Remember to assemble the plates and spacers before you mix the ingredients for the gel of polyacrylamide because the solution must be poured into the mold immediately after adding the catalysts.

Prepare the separating gel (30 min).

- Mix the following ingredients for the 15% separating gel:

30% acrylamide, 0.8% bisacrylamide	15 mL
1.5 M Tris chloride, 0.4% sodium dodecyl sulfate, pH 8.8	7.5 mL
Water	7.1 mL
10% (m/v) ammonium persulfate	0.40 mL
Tetramethylethylenediamine	0.015 mL
Total	**30 mL**

- Pour the mixture into the mold and flatten the top as you did before.

Prepare samples of proteins for electrophoresis (30 min).

TABLE 2.1   Aliquots of Protein for Electrophoresis[a]

	Number of Cells Producing Each Milliliter of Lysate	Concentration of Protein (mg mL^{-1})	Volume (μL)	Mass (μg)
Standard polypeptides[b]	NA	2	8	16
Uninduced lysate	~5 × 10^6			25
Induced initial supernate	~5 × 10^6			25
				50
LuxG-hexahistidine purified by adsorption	NA			2
				10

[a] *The concentrations of protein from the Bradford assays will allow you to complete this table.*
[b] *The standards from Sigma-Aldrich have already been unfolded and saturated with dodecyl sulfate. This sample is ready to load and does not need to be heated.*

The mass of protein from each sample to be loaded is shown in Table 2.1.

- Add in the proper order (Table 1.3) the appropriate amounts of water, buffer, and 2-sulfanyl ethanol. Calculate to be sure that at least 2 μg of sodium dodecyl sulfate will be added for every microgram of protein in addition to the amount of sodium dodecyl sulfate needed to make its final concentration in the sample 0.1%, and then add the dodecyl sulfate.

- Immediately heat the sample in the boiling water for 1 min, then add the Bromophenol Blue and the glycerol. Again, the final volume of each sample should be 50 μL. After they are prepared, the samples can sit for as long as necessary at room temperature.

Pour the stacking gel (20 min).

- Mix the ingredients for the 4.5% stacking gel, pour the mixture into the mold, and insert the comb as in FNR Exercise 4.

Electrophoresis (2 h).

- Fill the reservoirs of the electrophoretic apparatus with running buffer as in FNR Exercise 4.

⁎ Run at 125 V (about 55 mA) until the dye enters the separating gel (about 30 min).

⁎ Increase the voltage to 250 V (about 110 mA). Run until the dye is near the bottom (about 60 min).

⁎ The gel should run for a total of about 300 volt-hours.

Staining and destaining (30 min of laboratory session followed by an overnight step).

⁎ Stain the gel with Simply Blue and NaCl at room temperature overnight as described in FNR Exercise 4. As before, if one is available, it is best to use a plastic tray that has a lid to prevent evaporation. Alternatively, Saran Wrap or aluminum foil may be used. The quick method using the microwave oven described earlier may not be effective in this case because of the higher concentration of acrylamide in the gel.

⁎ Make a photographic image of the gel as described in FNR Exercise 4.

Analysis of the Data and Questions to Address in Your Report

1. Prepare a plot of the logarithm of the relative mobilities (R_f) of the standard polypeptides as a function of their length, as explained in FNR Exercise 4.

2. Estimate the length of the polypeptide constituting LuxG-hexahistidine by interpolation on the plot of the logarithms of the relative mobilities. Is your estimate close to the actual length of the polypeptide?

3. Compare the pattern of the bands in the lanes of the uninduced lysate and the induced lysate. What is the evidence that the induction worked or did not work?

4. Comment on the effectiveness of the purification by adsorption and elution from the solid phase to which Ni^{2+} had been chelated.

5. Does this experiment provide any clues as to the quaternary structure of LuxG from *P. leiognathi*?

6. Is it possible to assay the activity of an enzyme after this type of electrophoresis?

7. What is the consequence of using 15% polyacrylamide in the separating gel, rather than 12.5% as in the similar exercises for FNR and LDH?

LuxG EXERCISE 4

Assay for the Flavin Reductase Activity of LuxG-hexahistidine

Technical Background

There are several parameters that contribute to the design of the assay for the enzymatic activity of LuxG. The absorption maximum (Figure 2.6A) of NADH at 25 °C is at 339 nm (McComb et al., 1976; Rahman et al., 2010). At 339 nm, the extinction coefficient of NADH is 6310 M^{-1} cm^{-1} (Bergmeyer, 1975; McComb et al., 1976). At pH 8.0, there is an isosbestic point at 336 nm in the spectra (Figure 2.6B) for the conversion of FMN to $FMNH^-$ (Song et al., 2007a). Because of this isosbestic point, at 336 nm, the absorption of a solution resulting from the FMN and the $FMNH^-$ that it contains does not change as either is converted into the other. At 336 nm the extinction coefficient for NADH is 6280 M^{-1} cm^{-1} (McComb et al., 1976). The extinction coefficient of NAD^+ at 336 nm is 50 M^{-1} cm^{-1} (Rahman et al., 2010). Consequently, as NADH is converted to NAD^+ and FMN is converted to $FMNH^-$, the absorbance at 336 nm of the solution in a cuvette of 1-cm path-length should decrease by 0.0062 for every decrease in the concentration of NADH of 1 μM ($\Delta\varepsilon_{336}$ = 6230 M^{-1} cm^{-1}). The extinction coefficient of FMN at 446 nm is 12,200 M^{-1} cm^{-1} (Sucharitakul et al., 2007) and the extinction coefficient of $FMNH^-$ at 446 nm is 700 M^{-1} cm^{-1} (Song et al., 2007a). The extinction coefficients of NADH and NAD^+ at 446 nm are negligible. Consequently, as FMN is converted to $FMNH^-$ and NADH is converted to NAD^+, the absorbance at 446 nm of the solution in a cuvette of 1-cm path-length should decrease by 0.0115 for every decrease in the concentration of FMN of 1 μM ($\Delta\varepsilon_{446}$ = 11,500 M^{-1} cm^{-1}).

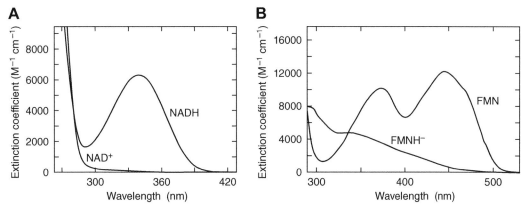

FIGURE 2.6　**Optical spectra of the substrates for LuxG.**　(A) Spectra for NADH and NAD^+ (Rahman et al., 2010) scaled to units of the extinction coefficient of NADH (Bergmeyer, 1975; McComb et al., 1976). (B) Spectra for FMN and $FMNH^-$ (Song et al., 2007a) scaled to units of the extinction coefficient of FMN (Sucharitakul et al., 2007).

In the following exercises, the enzymatic activity of LuxG will be quantified by following the decreases in A_{336} and A_{446}. Both the dA_{336}/dt and the dA_{446}/dt that occur when NADH is oxidized to NAD^+ and FMN is reduced to $FMNH^-$ can be converted directly into units of micromolar $minute^{-1}$. The FMN reductase (NADH) encoded by the *luxG* gene will also use riboflavin in place of FMN. Riboflavin has all of the same spectral properties as FMN, which is simply riboflavin 5′-phosphate.

Experimental Plan

Three samples will be examined: (1) a lysate of uninduced Tuner(DE3)/pGhis bacteria, (2) the initial supernate of the lysate from induced Tuner(DE3)/pGhis from LuxG Exercise 1, and (3)

LuxG-hexahistidine purified by adsorption and elution from the solid phase with chelated Ni^{2+} in LuxG Exercise 2. The uninduced control is necessary because the host cells of *E. coli* used for expression of the recombinant LuxG from *P. leiognathi* express a riboflavin reductase (NAD(P)H) of their own (Fieschi et al., 1995; Ingelman et al., 1999). The signal obtained for the uninduced sample (only host proteins) is subtracted from that for the induced sample (host and recombinant proteins) to estimate the enzymatic activity from the induced LuxG in the lysate. The endogenous riboflavin reductase (NAD(P)H) of *E. coli* will not be present in the preparation of purified LuxG-hexahistidine so no correction is required in this instance.

Reagents, Supplies, and Equipment

Buffered solution for the assay of FMN reductase (NADH): 10% glycerol, 1 mM dithiothreitol, 50 mM Tris chloride, pH 8.0.

Samples of enzyme:
A sample of uninduced Tuner(DE3)/pGhis lysate.
A sample of the initial supernate of the lysate from induced Tuner(DE3)/pGhis.
A sample of purified LuxG-hexahistidine.

Stocks of reactants:
20 mM NADH (minimize exposure to air to prevent oxidation).
4 mM FMN (FMN is more soluble than riboflavin because of its phosphoryl group).
400 μM riboflavin.

Supplies and equipment for spectrophotometry.
If you are using a plate reader, you will need a multiwell plate that is transparent to ultraviolet light for A_{336} assays and a standard plastic plate for A_{446} assays (Figure 1.13 and Appendix I).

If you are using a conventional spectrophotometer, you will need cuvettes that have a narrow, tapered chamber, for which only 0.5 mL of sample is required. The preference for the type of plastic for A_{336} and A_{446} assays is as for the multiwell plates.

If you are using a Spec 20, the glass culture tubes used with this device are partially transparent to light of 336 nm but fully transparent to light of 446 nm.

Before You Come to the Laboratory

The volume of the assay that you will use is determined by which type of spectrophotometer you will be using, 0.2 mL (multiwell plate), 0.5 mL (cuvette), or 3 mL (Spec 20 tube). Determine what this volume will be, then perform the following calculations:

- Calculate the molar mass of LuxG-hexahistidine in g mol^{-1}. This number is the same as the molar mass in μg μmol^{-1}.

- For the supernate from the lysate of the induced cells and the sample of LuxG-hexahistidine purified by adsorption and elution, estimate the concentration of LuxG-hexahistidine in micromolar. The induced initial supernate contains many proteins, but if you assume that the purification of the enzyme by adsorption had a 100% yield and all of the protein in that sample is LuxG-hexahistidine, you can estimate the amount of LuxG-hexahistidine that was in the initial mixture as follows:

1. First, calculate the number of micromoles of LuxG-hexahistidine you recovered from the adsorption and elution by dividing the number of micrograms of total protein that eluted by the molar mass of LuxG-hexahistidine in $\mu g \, \mu mol^{-1}$.

2. Assume that the number of micromoles of LuxG-hexahistidine calculated in the previous step was the amount in the sample loaded onto the column, and is thus close to the number of micromoles of LuxG-hexahistidine that was in the supernate of the lysate.

3. Using the volume of the supernate from the lysate, estimate the micromolar concentration of LuxG-hexahistidine in this supernate.

• Calculate the volumes of the supernate from the lysate of the cells and the volume of the solution of the purified enzyme that will give a final concentration of $0.4 \, \mu M$ LuxG-hexahistidine in each assay for enzymatic activity. You should not pipette any volume less than $10 \, \mu L$, in order to avoid the uncertainty of measuring small volumes, so you may need to dilute a sample from one or the other of the solutions of enzyme before taking the sample that you need for the enzymatic assay.

• No LuxG-hexahistidine should be present in the lysate of uninduced cells of Tuner(DE3)/pGhis. For the assays of this supernate, use a sample with the same mass of total protein as the one you will be using for the induced initial supernate.

• Refer to Table 2.2 regarding the various components to be added to the reactions. Calculate the volume of each component needed for the desired reaction volume. Because the volume of the samples of protein to be added will vary, the volume of the buffered solution for the assay of activity will also vary in order to reach the appropriate final volume for each reaction.

You should remember to correct for any dilutions you have to make to get the proper concentrations of enzyme in the assay when you are calculating the enzymatic activities after you have made the measurements.

TABLE 2.2   Final Concentration of Components in the Enzymatic Assay

Components to Add in the Order Listed	FMN Reaction	Riboflavin Reaction
Tris chloride, pH 8.0	~50 mM	~50 mM
FMN	80 μM	
Riboflavin		80 μM
NADH	200 μM	200 μM
LuxG-hexahistidine (as the pure protein or a component of lysate)	0.4 μM	0.4 μM

Procedure (One Laboratory Session, 3 h)

Prepare the mixtures for the assays (30 min).

Prepare assay mixtures that will contain the final concentrations listed in Table 2.2, after adding the samples of enzyme, by adding the ingredients in the order in which they appear in the table. First, add the calculated volume of buffer to the well (0.2 mL), cuvette (0.5 mL), or Spec 20 tube (3.0 mL), then add the appropriate volume of each of the other components. The NADH should not be added until just before placing the tube in the spectrophotometer to minimize its exposure to air, but the enzyme should be the last ingredient to be added, just after the NADH. Before you came to the laboratory, you decided on the proper volumes of water and the various solutions to add to each of the assays. After all of the components are present, stir gently to mix them well but not so vigorously as to oxidize the NADH with oxygen. Assay the enzyme with FMN as the oxidant and then with riboflavin as the oxidant.

Initiate the reactions and collect the raw data (2.5 h).

Assay the purified LuxG-hexahistidine first, then the supernate from the lysate, and finally the negative control of the supernate from the lysate of uninduced cells.

- For each assay you perform today, monitor the A_{336} of the reaction immediately after adding the enzyme at intervals of 30 s for at least 3 min. The initial A_{336} should be about 1.2, and the rate of decrease in A_{336} should be between 0.1 and 0.2 min^{-1}. If necessary, change the amount of enzyme you add to the assay to get the appropriate change in absorbance.

- Once you have decided on the proper amount of enzyme to add to the assay, immediately repeat each assay you do today just as you did the first assay, except this time monitor the A_{446} of the reaction immediately after adding the enzyme at intervals of 30 s for at least 3 min. The initial A_{446} should be about 1.0, and the rate of decrease in A_{446} should be between 0.2 and 0.3 min^{-1}.

- In addition to samples of uninduced Tuner(DE3)/pGhis lysate, which are included as controls for the background activity of riboflavin reductase (NAD(P)H) in the uninduced bacteria, you should also run controls in which no protein is added to the assay. This is to reveal the rate of oxidation of NADH by any oxygen in the sample, on the one hand, and any uncatalyzed reduction of FMN or riboflavin, on the other. Controls with no protein should be performed for each of the two flavins being used as reactants, at each of the two wavelengths.

Negative controls.

Any decrease in A_{336} or A_{446} observed in the enzymatic assays of the uninduced Tuner(DE3)/pGhis lysate will be subtracted from the respective decreases observed in the assays of the supernate of the induced lysate.

Any decreases in A_{336} or A_{446} observed in the enzymatic assays run in the absence of any protein will be subtracted from the respective decreases observed in the assays of the purified LuxG-hexahistidine, but not from the respective decreases for the supernate of the induced lysate because these uncatalyzed changes are already corrected for with the decreases for the supernate of the uninduced lysate.

Analysis of the Data and Questions to Address in Your Report

1. Calculate the concentrations of enzymatic activity (μmol min^{-1} mL^{-1}) for the LuxG-hexahistidine in the supernate of the lysate of the induced Tuner(DE3)/pGhis bacteria and the LuxG-hexahistidine that was purified by adsorption and elution. For those who require equations, the Beer–Lambert Law that can be used for this calculation was discussed in FNR Exercise 2. Remember that the milliliters in the calculation are the milliliters of the undiluted supernate and the undiluted solution of purified LuxG-hexahistidine that ended up in the assay. Account for any dilutions you had to make.

Perform the calculation for both the assays with FMN and with riboflavin, and for the measurements you made at 336 nm and at 446 nm. The absorbance at 336 nm should decrease by 0.0062 for every decrease in the concentration of NADH of 1 μM, and the absorbance at 446 nm should decrease by 0.0115 for every decrease in the concentration of FMN of 1 μM.

a. Plot the data and calculate the dA/dt for each of the 16 relevant reactions.

b. Subtract the appropriate rates of the background samples from the assays for LuxG-hexahistidine.

c. Convert each dA/dt to the concentration of enzymatic activity in micromoles minute^{-1} milliliter^{-1}

d. From the four concentrations of enzymatic activity, those for both riboflavin and FMN at both 336 and 446 nm, for each of the two samples, the lysate and the purified enzyme, estimate what the actual concentration of enzymatic activity is for each of the two samples.

2. From these two concentrations of enzymatic activity, calculate the specific activity (μmol min^{-1} mg^{-1}) for each of the two samples: LuxG-hexahistidine in the supernate from the lysate of the induced Tuner(DE3)/pGhis and the LuxG-hexahistidine purified by adsorption and elution.

3. Calculate the catalytic constant for LuxG-hexahistidine. Use the data for the purified LuxG-hexahistidine for this calculation. Although LuxG-hexahistidine functions as a homodimer, each subunit of the enzyme has an active site. The catalytic constant is the first-order rate constant for the rate at which one active site converts reactant to product when it is saturated with both reactants. Consequently, it is only the molar concentration of subunits, or polypeptides, of LuxG in a solution that is relevant to the calculation.

Assume that a concentration of FMN or riboflavin of 80 μM and a concentration of NADH of 200 μM are sufficient to saturate the active site so that at these concentrations of reactants the enzyme is running at full speed. The only molar concentration of active sites that you know with any accuracy is that for the purified enzyme, which you calculated before coming to the laboratory. Divide the concentration of enzymatic activity (μmol min^{-1} mL^{-1}) by the concentration of active sites (μmol mL^{-1}). The catalytic constant is always reported in units of second^{-1}.

4. Compare the catalytic constant (also called the turnover number) you have obtained to those reported in the literature for LuxG and other flavin reductases. Suggest explanations for any significant differences.

5. Discuss whether or not the calculated catalytic constant convinces you that the purified enzyme is in fact LuxG-hexahistidine.

TABLE 2.3   Summary for the Purification of Recombinant LuxG-hexahistidine

	Total Volume (mL)	Total Protein (mg)	Concentration of Protein (mg mL⁻¹)	Total Activity (μmol min⁻¹)	Concentration of Activity (μmol min⁻¹ mL⁻¹)	Specific Activity (μmol min⁻¹ mg⁻¹)	Fold Purification	Recovery (%)
Initial supernate of induced Tuner(DE3)/pGhis								
LuxG-hexahistidine purified by adsorption and elution								

6. Prepare a table summarizing the purification (Table 2.3). To determine the total volumes after each step of purification, correct for the removal of aliquots for the various analytical procedures. Aliquots were removed for assays of total concentration of protein and enzymatic activity, and for electrophoresis. Be aware of the fact that the values entered for total volume at each step will affect the values calculated for total protein, total activity, and percent recovery. You should correct for aliquots removed from the most recent step as well as earlier steps in the purification. For example, if you removed 5% from the initial supernate and 4% from the pool of fractions from affinity adsorption, do the following. Correct the volume of the initial supernate by dividing by 0.95. Correct the volume of the pool of fractions from the affinity adsorption by dividing by 0.95 and then by 0.96.

7. If you have already completed the exercises for FNR or LDH or both of them, consider the following issue. When it works optimally, using a histidine tag and adsorption on a solid phase to which Ni²⁺ has been chelated allows for purification of the target protein in a single step. In contrast, purification of proteins from their natural source typically requires several steps. Purification by affinity adsorption is used for the purification of both FNR and LDH, but in both cases another method is used first to enrich the desired enzyme. Adsorption by anion exchange is used for FNR, while fractionation with ammonium sulfate is used for LDH.

Compare the tables summarizing the purifications you have performed and comment on the effectiveness of using a histidine tag and adsorption on a solid phase to which Ni²⁺ has been chelated relative to the other two techniques just mentioned. Is the time spent isolating a segment of DNA encoding the protein, inserting it into a plasmid so that a histidine tag is attached to the carboxy-terminus of the enzyme, and expressing the enzyme from the plasmid in bacterial cells to facilitate purification justified? Another point you should consider is that before an enzyme is purified from natural sources, its sequence of amino acids and hence the sequence of DNA encoding it, is usually unknown.

LuxG EXERCISE 5

Determination of the Michaelis Constant of LuxG-hexahistidine for NADH

Technical Background Experimental Plan

Refer to FNR Exercise 5. There are the following differences in this instance:

1. Nicotinamide adenine dinucleotide (NADH) is used here rather than NADPH. The range for the concentrations of NADH will be from 6.25 to 200 μM.

2. The cosubstrate is FMN rather than INT. The reported K_m of LuxG for FMN is 2.7 μM (Nijvipakul et al., 2008), so all reactions will have 80 μM of FMN to ensure that this reactant is at a concentration that saturates the enzyme as the concentration of NADH is systematically raised to reach the limiting velocity V (Equation i.2).

3. In this instance, the absorbance of FMN at 446 nm will be used to monitor the rate of the enzymatic reaction rather than the absorbance of NADH because the change in absorbance is greater (0.0115 rather than 0.0062 for every decrease in concentration 1 μM) and it is within the visible range. Even though the concentration of NADH is changing as the reaction progresses, its decrease in concentration during the reaction is always equal to the decrease in concentration of FMN at all concentrations because of the stoichiometry of the reaction. As you have already learned, its decrease in absorbance could also be used to assay the enzyme.

4. The decrease in A_{446} will be measured. The extinction coefficient of FMN at 446 nm is 12,200 M^{-1} cm^{-1} (Sucharitakul et al., 2007) and the extinction coefficient of $FMNH^-$ at 446 nm is 700 M^{-1} cm^{-1} (Figure 2.6B). Thus the change in absorbance should be 0.0115 for every decrease in the concentration of FMN, and stoichiometric increase in the concentration of $FMNH^-$, of 1 μM ($\Delta\varepsilon_{446}$ = 11,500 M^{-1} cm^{-1}).

Reagents, Supplies, and Equipment

Buffered solution for the assay of FMN reductase (NADH): 10% glycerol, 1 mM dithiothreitol, 50 mM Tris chloride, pH 8.0.

Solutions of reactants:
20 mM NADH (minimize exposure to air to prevent oxidation).
4 mM FMN.

Supplies and equipment for spectrophotometry.
Because the only wavelength to be monitored in this experiment is in the visible range, if plastic vessels are used, it is best that they be made of plastic made for measuring absorption of visible light. Plastic that is transparent to ultraviolet light may block some of the visible light.
If you are using a plate reader, you will need a plastic multiwell plate (Figure 1.13 and Appendix I).
If you are using a conventional spectrophotometer, you will need a cuvette (or cuvettes) that has a narrow, tapered chamber, for which only 0.5 mL of sample is required.
If you are using a Spec 20, its glass culture tubes are fully transparent to light of 446 nm.

A plate reader, conventional spectrophotomer or a Spec 20.

Before You Come to the Laboratory

- A dA_{446}/dt of about 0.15 min^{-1} is appropriate for the fastest reaction, which is the reaction with the greatest concentration of NADH. This concentration happens to be the concentration you used in the assays you performed in the last exercise. From your results of that exercise, decide how the purified LuxG-hexahistidine should be diluted and what volume should be added to each assay. In this instance, because you will be running assays in which the concentration of enzyme must be identical in each assay, plan your experiment so that you are adding a volume of 25 or 50 μL of enzyme to initiate each assay to ensure that the deliveries are precise. Again, the volume of the assays is determined by the type of spectrophotometer you will be using. The molar concentration of purified LuxG-hexahistidine used in the reactions will be needed for the calculation of the K_m after the data has been collected, so calculate that number now.

- Determine how you will dilute the stock solution of NADH and what volumes of each of the dilutions you will use to obtain the required concentrations for estimating the Michaelis constant. You will perform assays with concentrations of NADH of 0, 6.25, 12.5, 25, 50, 100, and 200 μM. See the instructions for serial dilutions in FNR Exercise 5.

- Plan what stock solution you should prepare before you begin the assays so that all you have to do for each assay is to add a particular volume of that stock solution, a particular volume of the serially diluted NADH, and a particular volume of enzyme, in that order.

Procedure (One Laboratory Session, 2 h)

- Perform the assays as you did in the preceding exercise with 80 μM of FMN in all of them and with concentrations of NADH of 0, 6.25, 12.5, 25, 50, 100, and 200 μM in separate assays. You will not need to perform any controls other than the assay that has no NADH, since that assay will control for all of the adventitious decrease in the concentration of FMN that may result from processes other than the enzymatic activity of LuxG-hexahistidine, which cannot function in the absence of NADH.

Analysis of the Data and Questions to Address in Your Report

1. For each concentration of NADH, calculate the concentration of enzymatic activity (μmol min^{-1} mL^{-1}) from the dA_{446}/dt. Divide each of the resulting values by the molar concentration of the active sites of LuxG-hexahistidine (μmol mL^{-1}) in each assay. Then convert minutes to seconds to obtain the enzymatic activity in second^{-1}. This is the same calculation you performed in LuxG Exercise 4 to obtain the catalytic constant. In the current analysis, however, variation in the concentration of the reactant NADH generates a range of values in second^{-1} rather than a constant.

2. Determine the Michaelis constant for NADH, K_{mNADH}, for LuxG by plotting the concentration of enzymatic activity (s^{-1}) as a function of the concentration of NADH (μM) and fitting Equation i.2 to the data. You should convince yourself that regardless of which of the two types of units for the concentration of enzymatic activity are used for plotting the data, micromole minute^{-1} milliliter^{-1} or second^{-1}, the same K_m will be obtained in the same units, which are those units used for the concentration of reactant, which are micromolar in this

case. There are also Michaelis constants for FMN, K_{mFMN}; NAD$^+$, K_{mNAD}; and FMNH$^-$, K_{mFMNH}, that you could measure if you had the time to do so.

3. Discuss whether or not the K_{mNADH} that you have measured convinces you that the enzyme you have purified is in fact LuxG-hexahistidine.

4. What concentrations of NADH is LuxG likely to encounter in the cytoplasm of *P. leiognathi*?

5. Compare the value of K_{mNADH} for LuxG with the Michaelis constants that you have measured for FNR or LDH or both of them. What does this tell you regarding the concentration of the reduced form of NAD(P) these enzymes are likely to encounter in the cytoplasm of *Spinacia oleracea* or *Bos taurus*, respectively?

LuxG EXERCISE 6

Determination of the Molar Concentration of LuxG-hexahistidine Spectrophotometrically

Technical Background

When you are dealing with a solution of protein that is a mixture of many different enzymes and other proteins, the concentration of protein is always expressed in the unit of milligrams because the only unit appropriate to unanalyzed mixtures is mass. As a chemist, however, you already know that expressing the amount of any molecule in the unit of moles is far more informative than in the unit of grams. To measure the moles of a protein in a sample, however, the sample must contain only that protein. From now on, whenever possible, you should get in the habit of expressing the concentration of a particular enzyme in units of molarity, usually micromoles of active sites liter^{-1}, and the amount of any enzyme in the unit of nanomoles of active sites. This habit will pay off.

The molar concentration of the active sites of an enzyme can be determined by measuring the absorbance at 280 nm (A_{280}) of a solution containing only that enzyme if the sequence of amino acids in the mature enzyme is known. The aromatic side chains of the amino acids tryptophan, tyrosine, phenylalanine, and histidine absorb ultraviolet light, as do the disulfides in cystines (Edelhoch, 1967; Pace et al., 1995; Perkins, 1986; Wetlaufer, 1962). At a wavelength of 280 nm, only the indolyl groups of the tryptophans, the *p*-hydroxyphenyl groups of the tyrosines, and the disulfides of the cystines in a protein absorb strongly (Figure 2.7).

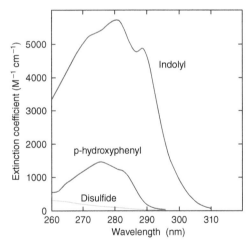

FIGURE 2.7   **Spectra of the side chains in the three amino acids that absorb ultraviolet light most strongly.**   The spectra are for the indolyl group (red) in the side chain of *N*-acetyltryptophanamide, the *p*-hydroxyphenyl group (blue) in the side chain of glycyl-L-tyrosylglycine, and the disulfide (yellow) in cystine (Edelhoch, 1967). The first two compounds are each an example of a model compound for the respective amino acid in a polypeptide, where it is attached at both sides by amido linkages to its neighbors. The ammonio group and the carboxylato group in a free amino acid have small but measurable effects on the absorption of ultraviolet light by the side chains. The phenyl group in phenylalanine and the imidazolyl group in histidine also absorb ultraviolet light, but their extinction coefficients are less than those of the disulfide in a cystine at all wavelengths, and are thus not included on the plot shown here. The measurements of the spectra were made at pH 6.5 to avoid the ionization of the *p*-hydroxyphenyl group in glycyl-L-tyrosylglycine (pKa = 9.8), which increases its absorbance at 280 nm by almost a factor of two.

The imidazole used to elute the LuxG-hexahistidine also absorbs at 280 nm and a solution containing 250 mM imidazole (the concentration present in the sample of LuxG-hexahistidine purified by adsorption and elution) will have a significant A_{280}, even though the imidazolyl groups in LuxG-hexahistidine, which are present at a much smaller molar concentration, will not.

Experimental Plan

To decrease the concentration of imidazole and dithiothreitol, which both absorb at 280 nm, in the LuxG-hexahistidine purified by affinity adsorption, two rounds of dilution and concentration by ultrafiltration under centrifugal force will be performed. Each dilution will decrease the concentrations of imidazole and dithiothreitol tenfold. After each centrifugation, the volume of the retentate should be about 10% of the volume before the centrifugation. Most of the protein will remain in the retentate above the filter while the imidazole will pass through in the filtrate. These two steps of dilution and concentration should return the protein to about its initial volume and initial concentration. The net effect of this procedure is to decrease the concentration of imidazole to around 2.5 mM and the concentration of dithiothreitol to 10 μM. The A_{280} of 2.5 mM imidazole and 10 μM dithiothreitol are both negligible, so the molar concentration of protein may now be determined by measuring the A_{280} of the solution.

Reagents, Supplies, and Equipment

All remaining samples of the LuxG-hexahistidine that you purified by affinity adsorption.

The buffered solution used for the assay of the enzymatic activity of LuxG without the dithiothreitol: 10% glycerol, 50 mM Tris chloride, pH 8.

Amicon Ultra-4 10K devices for ultrafiltration with capacities of 4 mL (Figure 1.16; you should need two).

Beckman J2 preparative centrifuge with JA-14 rotor (or equivalent), maintained at 4 °C.

Supplies and equipment for spectrophotometric assays:

If you are using an automated plate reader you need a multiwell plate transparent to ultraviolet light (Figure 1.13; refer to Appendix I regarding the use of multiwell plates and the plate reader).

If you are using a conventional spectrophotometer you need a cuvette that is transparent to ultraviolet light with a narrow, tapered chamber for a sample of 0.5 mL.

An automated plate reader or a conventional spectrophotometer that will assay absorbance at 280 nm.

Before You Come to the Laboratory

- Find the sequence of amino acids for LuxG, the FMN reductase (NADH) encoded by the *luxG* gene of *P. leiognathi*, in the UniProt database (http://www.uniprot.org).

- In the window above the sequence of amino acids, which says Blast as its default, select *ProtParam*. Without specifying an amino acid range, click Submit. By default, the composition of amino acids in the entire polypeptide should be shown.

- Using the following formula (Pace et al., 1995) for the molar extinction coefficient at 280 nm:

ε_{280} $(M^{-1}\,cm^{-1}) =$

$5{,}550\,M^{-1}cm^{-1}$ (number of Trp) $+\ 1{,}490\,M^{-1}cm^{-1}$ (number of Tyr) $+\ 125\,M^{-1}cm^{-1}$ (number of cystines) (2.3)
(Trp = tryptophan , Tyr = tyrosine)

and the number of tryptophans and tyrosines in the polypeptide constituting LuxG, calculate the molar extinction coefficient for a subunit of the enzyme. Cystines are almost never found in proteins that are expressed in the cytoplasm of a cell and that remain in the cytoplasm for their natural life. Because LuxG-hexahistidine was expressed in the cytoplasm of *E. coli*, and because LuxG is normally a cytoplasmic enzyme in *P. leiognathi*, it should not contain any cystines. Also, since dithiothreitol has been added to the purified LuxG-hexahistidine, any adventitious cystines formed during the purification should have been reduced back to cysteine.

Precaution

Do not use the centrifuge without supervision.

Procedure (One Laboratory Session, 1.5 h)

This exercise should be the last one that you perform with LuxG so you may now use all of the remaining enzyme left over from the preceding exercises.

- Thaw every remaining sample of purified LuxG-hexahistidine. Pool these samples and determine the total volume.

- Using 10% glycerol, 50 mM Tris chloride, pH 8, dilute this pool to 10 times its original volume.

- Add this solution of LuxG-hexahistidine to the upper chamber of the Amicon device (Figure 1.16) and spin it at 7000 rpm (7500 × *g*) in the JA-14 rotor for 10 min at 4 °C.

- Measure the volume of the sample of LuxG-hexahistidine. If it is greater than 0.4 mL, spin a bit more to bring the volume down.

- Dilute the concentrated sample tenfold, as before, then load into a fresh Amicon device (you should not reuse these devices).

- Spin as before to decrease the volume of the retentate to about 10% of the initial volume.

- If the volume of the sample of LuxG-hexahistidine is now equal to the volume of the original pool you made of all of the leftover LuxG, its concentration should be a little less than it was to begin with, and you have already estimated the molar concentration of enzyme in the original pool from its measured concentration of protein. If the current volume is different, calculate the new concentration. Ultraviolet spectrophotometry is most accurate in the range of absorbances from 0.1 to 1.0. If you anticipate that a dilution will be necessary to obtain an A_{280} in this range, make such a dilution with 10% glycerol, 50 mM Tris chloride, pH 8.

- Measure the absorbance of the solution of LuxG-hexahistidine at 280 nm.

- Measure the absorbance of the buffer alone at 280 nm.

Analysis of the Data and Questions to Address in Your Report

1. Subtract the absorbance of the buffer, if there is any, from the absorbance of the solution of enzyme, and use the ε_{280} of LuxG that you calculated before you came to the laboratory to determine the molar concentration of the LuxG-hexahistidine in the sample.

2. Compare the molar concentration of LuxG-hexahistidine obtained by direct spectrophotometry to that obtained by Bradford assay. If there is a discrepancy, suggest a possible explanation.

3. In theory, the best blank to use for the spectrophotometry would be the filtrate from the last ultrafiltration because it should be the solution in which the protein is dissolved without the protein. Why may it not be wise to use the filtrate as a blank? How could you convince yourself that the filtrate actually did not contain any protein?

Purification and Characterization of Bovine L-Lactate Dehydrogenase

BACKGROUND: IMPORTANCE OF L-LACTATE DEHYDROGENASE IN BACTERIAL AND EUKARYOTIC PHYSIOLOGY

L-Lactate dehydrogenase (LDH) catalyzes the oxidation–reduction:

$$pyruvate + NADH + H^+ \rightleftharpoons (S)\text{-lactate} + NAD^+ \tag{3.1}$$

In domestic cattle, there are five major isoenzymes of LDH. These five isoenzymes are formed by the random binomial combination of two isoenzymatic subunits, type A and type B, to form five different isoenzymatic tetramers (Figure 3.1). The A isoenzymatic subunit predominates in more anaero-

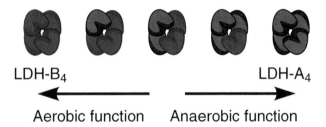

$$\text{LDH-B}_4 \qquad\qquad\qquad \text{LDH-A}_4$$

$$\longleftarrow \qquad\qquad \longrightarrow$$

Aerobic function   Anaerobic function

FIGURE 3.1 **The five isoenzymatic tetramers of LDH in vertebrates.** The isoenzymatic subunits formed from folded A polypeptides and the isoenzymatic subunits formed from the folded B polypeptides (Figure 3.3) combine binomially and randomly around axes of symmetry (Kyte, 2007, pp 466–498) to form the five hybrids represented in the drawing.

bic tissue such as skeletal muscle, while the B isoenzymatic subunit predominates in aerobic tissue such as heart (Berg et al., 2012; Chaikuad et al., 2005; Eszes et al., 1996; Everse et al., 1971; Everse and Kaplan, 1973; Holbrook, 1975; Voet and Voet, 2010; Voet et al., 2013). There are also several minor isoenzymes (Read et al., 2001) that will be of no concern in the following exercises. The variation in the distribution of the isoenzymes seems to respond to the demands of the tissues (Figure 3.2).

Genes encoding LDH have been characterized from the entire phylogenetic spectrum (Tsoi and Li, 1994). Sequences of amino acids for the human A and B isoenzymatic subunits of LDH, for the A and the B subunits for LDH from a plant, and LDH from a bacterium can be aligned (Figure 3.3). The sequences of amino acids for the bovine A and B isoenzymatic subunits that you will be studying are almost identical to the sequences of the human A subunit (93.7% identity) and B subunit (98.2% identity), respectively. In both cattle and humans, the polypeptide that folds to produce the A isoenzymatic subunit of LDH is 332 amino acids in length and that producing the B isoenzymatic subunit is 334 amino acids in length (Figure 3.3). The folded polypeptides then become subunits of the intact molecule of LDH. The holoenzyme of LDH is formed from

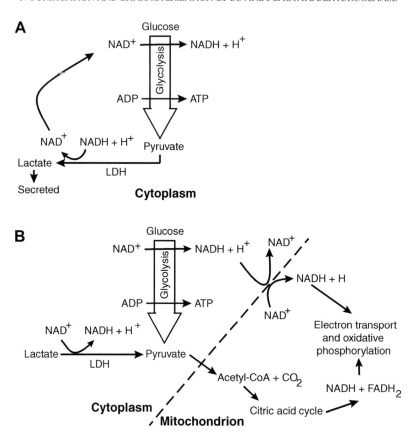

FIGURE 3.2   The physiological role of LDH in animal cells is affected by the availability of O_2.   As is necessarily the case for the reactions catalyzed by all enzymes, LDH catalyzes the conversion between pyruvate and NADH and lactate and NAD^+ in either direction. Oxidative phosphorylation is powered by electron transport in the mitochondria. Electron transport requires O_2, and the availability of O_2 in a tissue is one of the factors that determines the net direction of the catalysis by LDH in that tissue. (A) In skeletal muscle, the predominant isoenzymatic subunit of LDH is the A subunit. During short bursts of intense activity, insufficient O_2 gets to the cells in skeletal muscle, so electron transport in the mitochondria is slow. A great deal of pyruvate, however, is being generated by the anaerobic glycolysis required to provide the ATP for the contraction of the muscle. Luckily, the pyruvate generated by glycolysis does not enter the mitochondria because the citric acid cycle is also slow because of the lack of O_2. Instead, this large flux of pyruvate being produced by glycolysis is reduced to lactate by LDH. When the tissue is anaerobic, the sole purpose for this active reduction of pyruvate is not to produce lactate but to oxidize NADH formed by glycolysis to produce the NAD^+ that is necessary to sustain glycolysis. The lactate, which is a byproduct of this regeneration of NAD^+, is secreted because it has no other value to the cells of the muscle. During extended activity—for example, in sports requiring endurance—O_2 gradually becomes available to the cells of the muscle, pyruvate begins to enter the mitochondria, and the net conversion of NADH to NAD^+ catalyzed by LDH decreases. In these circumstances, the citric acid cycle can regenerate much of the NAD^+ needed to sustain glycolysis. (B) In the adult heart, the B isoenzymatic subunit predominates. In this tissue, because it must contract and relax continuously, O_2 is continuously available to the mitochondria. When O_2 is plentiful, pyruvate is transferred into the mitochondria as soon as it is produced by glycolysis to feed the citric acid cycle, and any lactate that enters the cell from the blood is converted by LDH to pyruvate to join the stream. It is critical that the heart have a constant supply of pyruvate to feed the citric acid cycle, so cardiac cells also import lactate and LDH oxidizes that imported lactate to pyruvate also. The NAD^+ needed to keep glycolysis running is provided by the electron transport that is continuously producing NAD^+ in the mitochondrion. These oxidizing equivalents of mitochondrial NAD^+ are relayed to the cytoplasm by the malate–aspartate shuttle (Voet and Voet, 2010). If the animal is lucky, LDH rarely has to supply cytoplasmic NAD^+ for glycolysis in cardiac cells, but it has to do so during congestive heart failure, when the supply of O_2 is cut off.

assembly of four of these subunits. Five isoenzymatic hybrids, two homotetramers and three heterotetramers, are possible: A_4, A_3B, A_2B_2, AB_3, and B_4 (Figure 3.1). The relative abundance of these isoenzymes varies in an organ-specific manner. The differences in charge among these bovine isoenzymes are significant enough that they can be distinguished electrophoretically. In the case of LDH, where the two different isoenzymatic subunits can associate at random in the tetramer to produce five isoenzymatic hybrids, it is only the subunits that differ in their kinetic properties. In the hybrid, each subunit functions independently, and their assembly into a tetramer is only an experimental distraction of little or no functional significance.

The enzyme from *Lactobacillus delbrueckii* is 32.2% identical to the bovine A isoenzymatic subunit (Figure 3.3). This bacterium is the species that produces the lactic acid during the fermentation

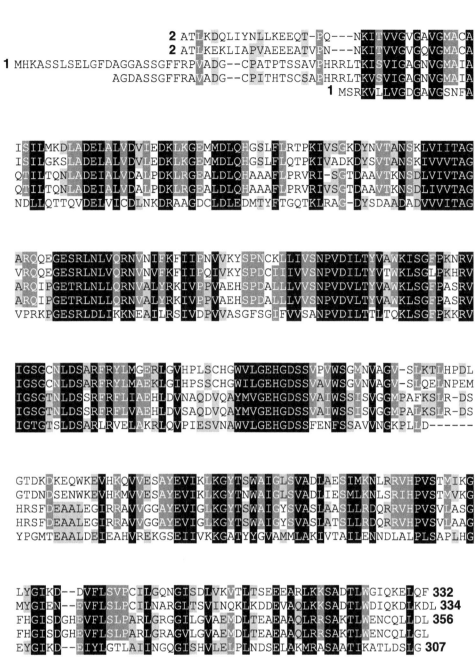

A

FIGURE 3.3 **Alignments of the sequences of amino acids for LDH from different species and for different isoenzymatic subunits of LDH, and an analysis of the respective secondary structures.** (A) Alignment of the sequences of amino acids for LDH from an animal (*Homo sapiens* A and B isoenzymatic polypeptides) (Read et al., 2001), a plant (*Hordeum vulgare* (barley) A and B isoenzymatic polypeptides) (Hondred and Hanson, 1990), and a bacterium (*Lactobacillus delbrueckii*) (Wang et al., 2013). The alignment shows that the sequence for the enzyme has been well conserved by natural selection. Shading indicates similar or identical amino acids in all four (darkest), three (intermediate), or two (lightest) sequences. In both instances, the sequences for the respective isoenzymatic polypeptides from one species are more similar to each other than they are to those for LDH from the different species.

of milk to yogurt. Because the production of ethanol is the mechanism used by yeast to regenerate the NAD⁺ needed for glycolysis and because, historically, the production of ethanol has been given the name fermentation, the production of lactate for the purpose of regenerating NAD⁺ for glycolysis by LDH is also referred to as fermentation when it occurs in microorganisms such as *L. delbrueckii*. When milk is turned into yogurt, the *L. delbrueckii* are not intent on making yogurt, their intent is to regenerate the NAD⁺ needed for glycolysis. The lactic acid is simply a waste

B

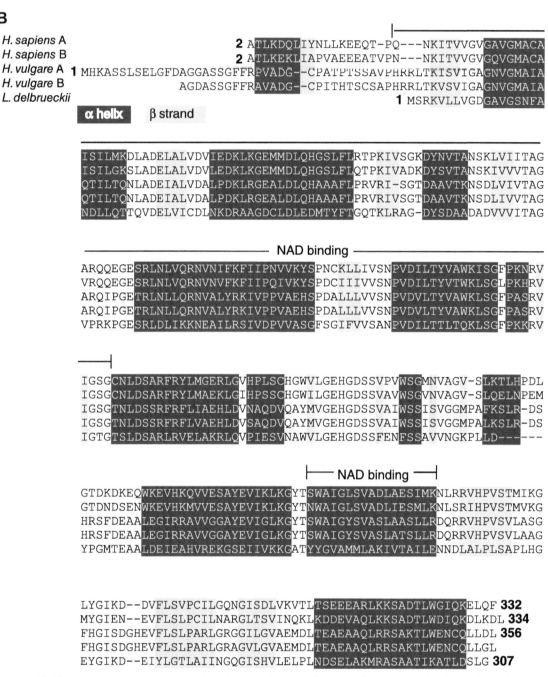

FIGURE 3.3 **Cont'd** (B) Secondary structure was identified by using the published crystallographic molecular model of the homotetramer of A subunits from *H. sapiens* (Read et al., 2001), entry 1I10 in the Protein Database of the Research Collaboratory for Structural Bioinformatics. The alignment, designation of secondary structure, and shading were performed with the Genedoc software described in the computational section. The investigators who submitted the crystallographic molecular model identified the portions of the polypeptide that form the Rossmann fold, a recurring domain of secondary structure that was one of the first of the many recurring domains that have been identified (White et al., 1976). As noted on the diagram, in lactate dehydrogenase, this domain and an α helix from the other domain in the molecule are responsible for binding NAD. In tabulations of the sequences of amino acids for polypeptides, the numbering of the amino acids starts with the initiator methionine at position 1. This methionine is known to be absent from the mature form of both isoenzymatic subunits of human LDH but is assumed to be present in the mature form of isoenzymatic subunit A from barley and the LDH from *L. delbrueckii*. No numbers are shown for isoenzymatic subunit B from barley because the complete sequence of this polypeptide is not yet available.

product as it is in skeletal muscle, even though it produces a desirable flavor and preserves the milk. The production of ethanol and the production of lactate are two of the many ways that organisms regenerate NAD^+ for anaerobic glycolysis by producing a reduced byproduct.

Barley has two genes encoding isoenzymatic subunits of LDH (Hondred and Hanson, 1990) and the sequences of amino acids for these polypeptides are more similar to each other than they are to vertebrate L-lactate dehydrogenases (Figure 3.3). As in mammals, these two types of polypeptides in barley assemble into five types of hybrid tetramers (Figure 3.1). The enzymatic activity of LDH in terrestrial plants is highest in the roots, where it produces lactate, again to regenerate the NAD^+ needed for anaerobic glycolysis. It, however, has also been detected at low levels in leaves from spinach (Sugiyama and Taniguchi, 1997).

The bovine LDH that you will be studying in these exercises is L-lactate dehydrogenase. Genes encoding D-lactate dehydrogenase (cytochrome), referred to here as D-LDH, have also been detected in the entire phylogenetic spectrum. Sequences for this enzyme have been found in bacteria (Taguchi and Ohta, 1991), fungi, plants (Engqvist et al., 2009), and animals (Flick and Konieczny, 2002). In eukaryotes, D-LDH appears to be localized to the mitochondria. L-LDH and D-LDH show no similarity in their sequences of amino acids, have polypeptides of significantly different lengths, and seem to be unrelated to each other.

SUMMARY OF THE EXERCISES WITH LDH

Exercises 1–3

Enzymatic Activity and Ionic Properties of the Isoenzymes of LDH in Crude Extracts of Bovine Tissues

These exercises investigate the variation in the isoenzymatic composition and enzymatic activity of LDH from bovine heart, kidney, liver, and skeletal muscle. The ionic properties of the isoenzymes from these tissues will be compared using electrophoresis on cellulose acetate.

Exercises 4–6

Purification of LDH from Bovine Heart

These exercises demonstrate methods for purification of an enzyme from the tissue of an animal. Fractionation with ammonium sulfate separates proteins based on their solubility in concentrated solutions of ammonium sulfate and results in some enrichment of the desired enzyme, but it is also used to eliminate components in the cytoplasm of the cells other than proteins and to concentrate the protein. Affinity adsorption followed by affinity elution is then used to purify LDH in one step. The solid phase that will be used is one to which adenosine monophosphate (AMP) has been covalently attached. Because AMP corresponds to part of the NADH structure, it is thought to associate specifically with the active site of the enzyme. Many other enzymes, however, associate specifically with AMP, so in this instance it is the procedure for eluting LDH that is specific for this one enzyme. The elution of the bound LDH from the matrix is performed by using a bisubstrate adduct of pyruvate and NAD^+ that has a high affinity for LDH. Because the adduct competes with AMP for only the active site of LDH, only LDH dissociates from the solid phase and all of the other proteins bound to the attached AMP remain bound to the solid phase. The concentration of LDH in the sample at each stage of the purification will be quantified by measuring the concentration of enzymatic activity. The mass of protein at each step will be determined spectrophotometrically using the Bradford assay or direct ultraviolet absorbance.

Exercises 7, 8, and 10

Characterization of the Structural Properties of LDH Purified from Bovine Heart

These exercises allow one to perform three techniques that are frequently used for analytical biochemistry: electrophoresis of native proteins on a polyacrylamide gel, electrophoresis of unfolded polypeptides saturated with dodecyl sulfate on a polyacrylamide gel, and chromatography by molecular exclusion (gel filtration). The electrophoresis of native proteins on a polyacrylamide gel will be used to analyze the proteins in the supernate from the homogenate of the heart and the purified LDH. Coomassie Brilliant Blue will be used to stain proteins in general, and a stain specific for the enzymatic activity of LDH will be used to locate the enzyme itself. Electrophoresis of unfolded polypeptides saturated with dodecyl sulfate allows all of the polypeptides in a sample to be catalogued after they have been separated electrophoretically. Thus, it is a good assay for the purity of the enzyme that has been eluted with the bisubstrate adduct. Electrophoresis of unfolded polypeptides saturated with dodecyl sulfate also permits an estimate of the length of each polypeptide to be made based on its electrophoretic mobility. Chromatography by molecular exclusion will be used to estimate the size of native LDH and hence its quaternary structure.

Exercises 9 and 11

Assay of the Michaelis Constant, K_{mNADH}, for NADH and the Detection on Immunoblots of the Homotetrameric Isoenzymes of LDH in Skeletal Muscle and Heart

A series of assays for the enzymatic activity of the isoenzymes in the extract from bovine skeletal muscle from the first exercise and the LDH purified by affinity elution from heart will be performed at a saturating concentration of pyruvate but variable concentrations of NADH to determine the Michaelis constant, K_{mNADH}, of these isoenzymes of LDH for NADH. In addition to comparing the values of K_{mNADH} for the two homotetrameric isoenzymes of LDH that predominate in skeletal muscle and heart, respectively, these Michaelis constants can be compared to the Michaelis constants you obtained for FNR and LuxG. An immunoblot of the extract from bovine skeletal muscle and samples from the various stages of purification of LDH from bovine heart will be performed. Polypeptides are unfolded with dodecyl sulfate, separated by electrophoresis, transferred to a sheet of nitrocellulose, and the polypeptides of the isoenzymes of LDH will be detected by immunostaining the nitrocellulose with antibodies specific for the A polypeptide and the B polypeptide, respectively.

LDH EXERCISE 1

Preparation of the Isoenzymes of LDH from Extracts of Various Tissues

Technical Background

Homogenates will be prepared from four adult bovine tissues: heart, kidney, liver, and skeletal muscle. These tissues provide a variety of physiological situations in which the isoenzymes of LDH are required to function. For example, there is variability in the availability of oxygen to the cells in these different tissues. In addition, gluconeogenesis, the regeneration of glucose from lactate released into the circulation by tissues such as skeletal muscle that are undergoing anaerobic respiration, is more prevalent in liver than in the other three tissues (Berg et al., 2012; Champagne et al., 2005; Kietzmann, 2004; Lehninger et al., 2008; Tijerina, 2004; Voet and Voet, 2010). Another tissue-specific variation is that the oxidation of this circulating lactate by the citric acid cycle is more prevalent in the heart and liver than in the other tissues. These processes require that, in the heart and liver, more lactate is converted to pyruvate than pyruvate is converted to lactate, which is the reverse of the direction of this net conversion in skeletal muscle.

Experimental Plan

It is best if each pair of students (or each student) be assigned one of the four tissues. The tissue is sheared into small fragments in a kitchen blender. A motor-driven *Potter–Elvehjem homogenizer* is then used to shear the fragments of tissue further, break open the cells, and release soluble cytoplasmic enzymes such as LDH. Insoluble structures such as fragments of cellular membranes with their associated proteins and polysaccharides, subcellular organelles such as mitochondria, extracellular matrix, and fibrous material are removed by centrifugation. Aliquots of the supernate are then removed and quick-frozen for use in subsequent exercises.

Reagents, Supplies, and Equipment

Bovine tissues: heart, kidney, liver, and skeletal muscle chopped into pieces of approximately 3 cm and put through a meat grinder in the cold room. The grinding should be done just before class is to begin or after class has started.

Chilled 10 mM EDTA, 50 mM sodium phosphate, pH 7.5.

Bucket filled with ice.

Thermometer.

Reusable Oak Ridge tubes (Figure 1.6) with a volume of 38 mL made from opaque plastic with a thick (2–3 mm) wall, a screw cap, and a round bottom for high-speed ($26,000 \times g$) spins.

Meat grinder.

Kitchen blender (Figure 1.7).

Potter–Elvehjem homogenizer (Figure 3.4).

Beckman J2 preparative centrifuge and JA-14 rotor or equivalent.

Insulated Dewar flask containing a slurry of dry ice in isopropanol to freeze samples of each extract.

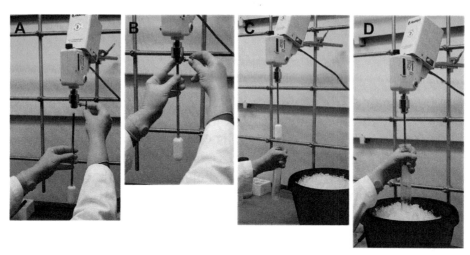

FIGURE 3.4 **Mammalian cells are released from fragmented tissue and sheared open with a motor-driven Potter–Elvehjem homogenizer.** (A) The apparatus is mounted on a wall rack or ring-stand in the cold room. The stem of the Teflon pestle is inserted into the *chuck* of the motor that will drive the pestle. (B) The *key for the chuck* is used to secure the pestle. Before it is used in the barrel of the glass tube, the motor should be turned on and the pestle checked for wobble. If it wobbles severely, the pestle could shatter the glass of the tube. Adjust the stem in the chuck to eliminate most of the wobble. (C) A batch of the blended tissue (50 mL) is poured into the glass tube of the homogenizer. The tube is gripped firmly near the top. It is very unlikely that the glass tube will shatter, but if this were to happen, it would probably crack near the bottom. (D) The glass tube is slowly raised so that the pestle gradually enters the blended suspension. Because of the tight clearance between the pestle and the wall of the glass tube as well as the high viscosity of the blended tissue, some force is necessary to push the pestle into the tube. Make sure that you push the pestle all the way to the end of the tube so that all of the suspension is sheared. As you do this, imagine the suspension flowing up through the tight space between the pestle and the tube where the shearing is occurring. The pestle is passed through the suspension of blended tissue three times. If practical, the tube is kept in ice during homogenization to minimize increase of temperature.

Precautions

Keep tissues and extracts chilled at all times, along with anything else that comes into contact with them. Use ice to chill the tubes, homogenizer, and glass container of the blender.

Wear safety glasses and disposable gloves when using the homogenizer; hold the homogenization cup firmly near its top, not at the bottom (Figure 3.4). Although it is unlikely that it will shatter, if this does happen it will probably be at the bottom.

Wear disposable gloves when handling tissue to protect the samples as well as yourself.

Do not use the centrifuge without supervision.

Procedure (One Laboratory Session)

Fragmentation of tissue and lysis of cells (1 h).

- You and your partner will be assigned a tissue: heart, kidney, liver, or skeletal muscle. Go to the cold room and weigh out about 40 g of the ground tissue in a weigh boat. You do not have to have exactly 40 g, but record the mass.

- Mix in a beaker the ground tissue with 4 mL of ice-cold 10 mM EDTA, 50 mM sodium phosphate, pH 7.5, for every gram of tissue. Bring the suspension to less than 5 °C on ice.

- Fragment the ground tissue in a blender (Figure 1.7). Blend at the highest speed for a total of 3 min for heart and skeletal muscle, 2 min for kidney, or 30 s for liver. Blend only 30 s at a time and check the temperature between each 30 s period. Keep the temperature below 20 °C. After you have blended the tissue, pour the suspension back into the beaker on ice.

- Stir the suspension well, and while it is still swirling, pour 50 mL into a graduated cylinder.

- Homogenize only this 50 mL portion in the Potter–Elvehjem homogenizer (Figure 3.4). The remainder will be discarded if all goes well. Don't allow the temperature of the homogenate to exceed 20 °C. This type of homogenization tears open mammalian cells with shearing force but nuclei and mitochondria remain intact (Boyer, 2000; de Duve et al., 1955).

Removal of insoluble material by centrifugation (40 min).

- Transfer the homogenate to two chilled and labeled Oak Ridge centrifuge tubes (Figure 1.6). Balance pairs of tubes on a double pan balance by adding the buffered solution for homogenization to the lighter one, and put the two tubes of a balanced pair into opposite holes in the rotor.

- Centrifuge the homogenates at 13,000 rpm (26,000 × g) for 20 min at 4 °C.

- Carefully pour the supernate into a beaker. Watch carefully that you do not pour out any of the pellet.

- Dispense aliquots of 2 mL of the supernate into each of five tubes, and put them on ice. If you homogenized skeletal muscle and the LDH exercise for determining K_{mNADH} is on the class schedule, prepare an extra 10 aliquots of 2 mL, for a total of 15.

- Most likely you will be performing the next exercise, the assay of the enzymatic activity of LDH, immediately after dispensing the aliquots. If so, leave them on ice until the assays are finished. If the assays will be performed in the next laboratory session, quick-freeze the aliquots as described below.

- If you are scheduled to do the assay for enzymatic activity today, perform the assay as outlined in the next section. Assay what is left after you have removed the aliquots rather than using up one of them.

Quick-freezing the aliquots of the extract of tissue for storage (20 min).

As before, the addition of glycerol to extracts of tissue before freezing them helps preserve enzymatic activity. Freezing the aliquot rapidly in a slurry of dry ice in isopropanol also helps preserve enzymatic activity. Add glycerol to all aliquots intended for freezing in this and subsequent exercises with LDH.

- It is imperative that you label each of the tubes with your initials, the date, and the type of tissue. There are two types of marking pens generally used in labs. The Sharpies are fine for assay tubes that are only needed for a single experiment, then discarded. For any tube that will be frozen in a slurry of dry ice in isopropanol you must use the alcohol-resistant Fisher or VWR laboratory markers! The isopropanol will dissolve the ink from other types of marker. Black ink sticks better than blue or other colors. Write directly on the tube; tape will fall off in the freezer.

- Add six drops of glycerol (three drops for every milliliter of homogenate) from a bottle-dropper or Pasteur pipette to each aliquot of homogenate. Each drop is around 30–40 µL, and the final concentration of glycerol should be around 10%. Mix the glycerol into each extract completely. Mix thoroughly by inversion or pipetting to disperse the glycerol. Avoid any vortexing that might cause foaming.

- Freeze in the slurry of dry ice in isopropanol in the insulated Dewar flask and store at −20 °C.

Planned use of the aliquots of the homogenate.
  Heart, kidney, and liver:
  - One for assay of enzymatic activity, two for electrophoresis of the isoenzymes, and two spares to cover for mistakes.
  Skeletal muscle:
  - One for assay of enzymatic activity, two for electrophoresis of the isoenzymes, ten for determination of K_{mNADH}, and two spares to cover for mistakes.

Analysis of the Data and Questions to Address in Your Report

1. Refer to FNR Exercise 1 and LuxG Exercise 1. Even if you have not performed these exercises, the sections in these exercises covering technical background, experimental plan, and procedures should help you to answer this question and the next. Consider the different methods used to lyse cells in the leaves of spinach, cells of *Escherichia coli*, and cells in mammalian tissue to release their cytoplasm. Discuss the structure of the tissues and cells and explain the reasons for the variation in technique.

2. Consider the centrifugations used to pellet chloroplasts and cells of *E. coli*. The homogenization of mammalian cells with a Potter–Elvehjem homogenizer does not disrupt nuclei or mitochondria (Boyer, 2000; de Duve et al., 1955). Discuss why the procedures specified for the FNR and LuxG exercises suggest that the subcellular organelles from these mammalian cells were pelleted during the centrifugation in the current exercise.

LDH EXERCISE 2

Spectrophotometric Assay of the Enzymatic Activity of Bovine LDH

Technical Background

If you have not already done so, you should read the Technical Background to LuxG Exercise 4. Lactate dehydrogenase catalyzes an equilibrium (Figure i.3 and Equation 3.1). The assay you will be using is designed so that, at early times, the reaction occurring in the assay is only the reduction of pyruvate to lactate by NADH rather than the oxidation of lactate to pyruvate by NAD^+, which is the reaction in the other direction of the equilibrium. Reduced nicotinamide adenine dinucleotide has a strong absorbance at 399 nm, A_{339}, but NAD^+ has almost none (Figure 2.6A), and lactate and pyruvate have none. At 339 nm, which is the maximum in its spectrum, the extinction coefficient of NADH is 6310 M^{-1} cm^{-1} (Bergmeyer, 1975; McComb et al., 1976) and the extinction coefficient of NAD^+ is 50 M^{-1} cm^{-1}, so the change in absorption when NADH is converted to NAD^+ has an extinction coefficient of 6260 M^{-1} cm^{-1} at 339 nm. Thus, the initial rate of the reaction, also referred to as the initial velocity of the reaction, in the direction dictated by the assay may be quantified by using spectrophotometry to measure the decrease in A_{339}.

The standard reaction mixture for the assay contains 0.4 mM NADH, 1 mM sodium pyruvate, 100 mM sodium phosphate, pH 7.5, and enzyme as needed (Eszes et al., 1996; Everse, 1971; Everse and Kaplan, 1973; Holbrook, 1975). The initial rate of the decrease in A_{339} (dA_{339}/dt) is converted into standard units for the concentration of enzymatic activity (μmol min^{-1} mL^{-1}).

Experimental Plan

The extract of the tissue you were assigned in the previous exercise will be assayed. The assay requires that you establish the optimal amount of extract to add to each reaction. To do this, several dilutions of extract are tested to determine which gives a decrease in A_{339} of approximately 0.15 min^{-1} in the standard reaction mixture.

Reagents, Supplies, and Equipment

The extract from the bovine tissue that you made in the previous exercise: heart, kidney, liver, or skeletal muscle (on ice).

40 mM NADH.

100 mM sodium pyruvate.

100 mM sodium phosphate, pH 7.5.

500 mM sodium phosphate, pH 7.5.

Plastic snap-cap test tubes (each holds 1.5 mL).

Parafilm.

Assorted pipettes:
One each: 20 μL Pipetman; 200 μL Pipetman; and 1000 μL Pipetman.

Timer.

Vortex mixer.

3. LDH

Spectrophotometric supplies and equipment:

If you are using a plate reader, you will need a multiwell plate that is transparent to ultraviolet light (Figure 1.13 and Appendix I).

If you are using a conventional spectrophotometer, you will need plastic cuvettes that are transparent to ultraviolet light. They should have a narrow, tapered chamber (only 0.5 mL sample required).

If you are using a Spec 20, use glass culture tubes that are partially transparent to 339 nm light.

Automated plate reader (Figure 1.13 and Appendix I), a conventional spectrophotometer, or a Spec 20.

Precaution

Avoid leaving a stock tube of NADH open to the atmosphere any longer than necessary. Oxygen from the atmosphere will oxidize it to NAD^+.

Procedure (One Laboratory Session, 1 h)

These instructions are written assuming 3 mL reactions will be performed in glass culture tubes for a Spec 20. If 0.5 mL cuvettes will be used in a conventional spectrophotometer, or 0.2 mL reactions prepared for a multiwell plate, scale down volumes as appropriate.

Dilute your extract of bovine tissue.

- For a 1:10 dilution, mix 50 μL of tissue extract and 450 μL of 100 mM sodium phosphate, pH 7.5, and gently vortex to mix well. For a 1:100 dilution, add 50 μL of tissue extract to 4.95 mL of buffer. If it is too viscous to pipette, pass the undiluted extract through a 27 gauge needle attached to a 2.5 mL syringe and then gently vortex it. Each time before you remove a sample to add to the assay mix, vortex the diluted extract.

- Label each of the tubes, and keep them on ice while you prepare the assay solution and zero the spectrophotometer.

- Zero the spectrophotometer at 339 nm against 3 mL of water.

Prepare the assay mix.

- For five assays each with a final volume of 3 mL, mix in a test tube:
  3 mL of 500 mM sodium phosphate, pH 7.5 (at room temperature).
  150 μL of 40 mM NADH (on ice).
  150 μL of 100 mM sodium pyruvate (on ice).
  11.5 mL of water (at room temperature).

- For each individual assay, remove 2.95 mL, and add it to a glass culture tube for the Spec 20.

- Put the tube in the spectrophotometer, and check to see that the A_{339} is between 0.6 and 0.8.

Assay diluted samples.

- Add 50 μL of a diluted tissue extract, place a piece of Parafilm over the top, and invert the tube several times to mix its contents thoroughly.

- Monitor the decrease in A_{339} for 3 min at intervals of 30 s.

  A decrease of A_{339} of around 0.15 min^{-1} indicates that the correct volume of the diluted extract has been used. A rate that is twofold faster or slower can be corrected by using a different volume of the diluted extract. Greater differences will require increasing or decreasing the dilution factor.

- Once you have determined the correct volume of the appropriate dilution to use in the assay, measure the rate for the diluted extract of the tissue until at least three consecutive runs agree closely. Record the volume and dilution of the extract you added to these triplicated assays.

Analysis of the Data and Questions to Address in Your Report

1. Calculate the initial dA_{339}/dt, the initial velocity for the LDH-catalyzed reaction, as you did in FNR Exercise 2 and LuxG Exercise 4.

2. As you also did in FNR Exercise 2, convert the initial velocity to a concentration of enzymatic activity (μmol min^{-1} mL^{-1}) using an extinction coefficient of 6260 M^{-1} cm^{-1} for the change in A_{339}. Correct for the dilution as appropriate.

3. Reduced nicotinamide adenine dinucleotide has another maximum of absorbance at 260 nm. Why do you measure LDH activity at 339 nm rather than at 260 nm?

4. Why was dA_{339}/dt measured for only 3 min in this experiment?

5. The direction of the interconversion of pyruvate and lactate catalyzed by LDH *in vivo* depends on the physiological circumstances. Describe all of the changes in the procedure and evaluation of the data that would be necessary to quantify the initial velocity for the reverse of the reaction that was assayed in this exercise.

6. Lactate has a chiral center. Was pyruvate converted to (S)-lactate, (R)-lactate, or a racemic mixture in this experiment? Explain how you know.

LDH EXERCISE 3

Separation and Visualization of the Isoenzymes of LDH by Electrophoresis on Cellulose Acetate

Technical Background

The five isoenzymes of bovine LDH, which are tetrameric hybrids of the isoenzymatic A and B subunits (Figure 3.1), can be separated by electrophoresis because each of the two subunits, and hence all five of the hybrids, have distinct titration curves (Figure 1.5) and hence different charges at the pH and ionic composition of the buffered solution in which the electrophoresis is run. The titration curve, isoionic point, and isoelectric point of a protein were discussed in FNR Exercise 1. If a protein is in a buffered salt solution with a pH equal to its pI, the net charge on the protein is zero. If the pH is below the pI, the protein will have positive charge. If the pH is above the pI, the protein will have negative charge. The charge on the protein will determine the magnitude of the electrostatic force felt by the protein in an electric field. The electrostatic force and the sieving of the medium in which the electrophoresis is run will determine the direction and the rate of migration of a particular protein in an electrophoretic separation (Berg et al., 2012; Boyer, 2012; Kyte, 2007, pp 36–45; Lehninger et al., 2008; Malmstrom et al., 2006). Since the five hybrids of the A and B subunits all have the same surface area, the sieving by the cellulose acetate is the same for all five. Consequently, they are separated only by differences in their charge, a fact that explains why the hybrids are so evenly spaced upon their separation by electrophoresis.

Experimental Plan

Samples of extracts of tissue or a purified enzyme are applied in the center of individual strips of cellulose acetate (Figure 3.5). The strip is carefully wetted with a buffered solution. The electrophoretic chamber has reservoirs for the anode and the cathode, each filled with the same buffered solution used to wet the cellulose acetate (Monthony et al., 1978). In this instance, because no stacking is used, which requires stable boundaries between two different solutions, the solution

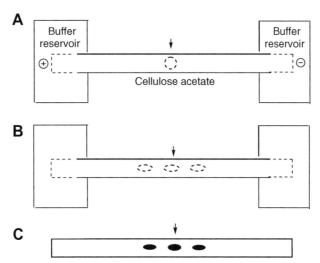

FIGURE 3.5 **Electrophoretic separation of isoenzymes on the basis of differences in their titration curves.** (A) A strip of cellulose acetate, wetted with buffered solution for the electrophoresis, is the polymeric network through which the molecules of protein move. The anode (left) and cathode (right) consist of positive and negative electrodes, respectively, submerged in reservoirs containing the buffered solution for the electrophoresis. A sample of an isoenzymatic mixture is applied to the center of the strip. The proteins are not visible at this point. (B) Isoenzymes with a net negative charge migrate towards the anode, those with a net positive charge towards the cathode, and those that are almost neutral remain near the center. During the electrophoresis, most of the electrical resistance in the circuit is in the cellulose acetate rather than in the reservoirs. Thus, the gradient of voltage equals the voltage supplied by the power supply divided by the length of the strip that is above the surfaces of the buffered solutions in the reservoirs. (C) The extent of migration of every protein that has enzymatic activity of interest is determined when the strip of cellulose acetate is stained (Figure 3.6).

throughout the system is the same. Voltage is applied across the strips in order to separate the proteins in each sample (Barnett, 1964; Mager et al., 1966).

The isoenzymes of LDH are visualized by a stain for their enzymatic activity that produces a color on the strip at the locations of the isoenzymes. The stain is a mixture of lithium lactate, NAD⁺, 5-methylphenazinium methyl sulfate (sold commercially under the peculiar name phenazine methosulfate), and Nitro Blue Tetrazolium (Opher et al., 1966). During the reaction, the enzyme is catalyzing the equilibrium in the reverse direction from the direction used for the spectrophotometric assay of enzymatic activity because the only reactants for the enzymatic reaction that are added are NAD⁺ and lactate (Figure 3.6). In the presence of only lactate and NAD⁺, NADH and pyruvate are formed. The concentrations of lactate and NAD⁺ in the staining reaction are much

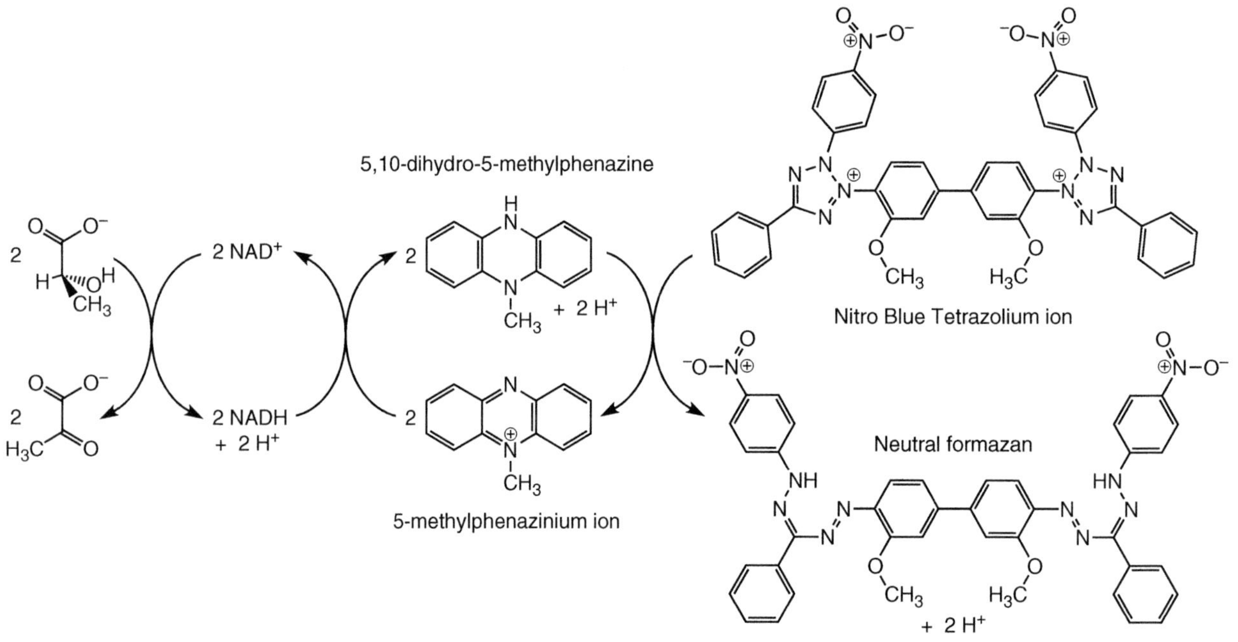

FIGURE 3.6 **The series of oxidations and reductions, ending with a chromogenic compound, that is used to stain a strip of cellulose acetate for the activity of LDH.** As LDH catalyzes the oxidation of lactate to pyruvate, a formal hydride ion is transferred on the active site of the enzyme from a molecule of lactate to a molecule of NAD⁺ to produce a molecule of NADH, which then dissociates from the active site. A formal hydride ion is transferred in solution to a 5-methylphenazinium ion from a molecule of NADH that has dissociated from an active site of LDH. A formal hydride ion is transferred in solution from a molecule of 5,10-dihydro-5-methylphenazine to one of the two tetrazolium rings in an ion of Nitro Blue Tetrazolium. Each of the two tetrazolium rings in a symmetrical Nitro Blue Tetrazolium ion can accept a hydride ion when being reduced. Each is reduced independently of the other, but when both are reduced, the fully reduced neutral formazan of the Nitro Blue Tetrazolium is formed. The dicationic Nitro Blue Tetrazolium ion is soluble in water, but the neutral formazan is insoluble, has a strong purplish blue color, and precipitates where it is formed to reveal the location of the various isoenzymes of LDH. The scheme is shown with a 2:2:2:1 stoichiometry for pyruvate:NADH:5-methylphenazinium ion:formazan because each Nitro Blue Tetrazolium ion accepts two formal hydride ions upon complete reduction. This method of staining can be used for detection of LDH on cellulose acetate and also in a polyacrylamide gel, in each case following the respective electrophoretic separation.

higher than the concentrations of pyruvate and NADH used in the spectrophotometric assay for the enzymatic activity of LDH in the previous exercise. These high concentrations are necessary to maintain saturation of the enzyme even when the local concentrations of the two reactants are rapidly decreasing in the vicinity of the bands of LDH. As the NADH is formed, it is reoxidized to NAD⁺ while it reduces the 5-methylphenazinium ion. The resulting 5,10-dihydro-5-methylphenazine reduces the Nitro Blue Tetrazolium ion to produce the neutral formazan, which, because it is no longer cationic, is insoluble and hence immobile. This reaction takes place wherever an isoenzyme of LDH is present, producing a blue band at its location on the strip. The number, locations, and intensities of the blue bands on the strip of cellulose acetate for one of the tissues provide an indication of the amounts of each of the isoenzymatic hybrids in that particular tissue. Over the four tissues that you homogenized, all five of the tetrameric hybrids of the isoenzymatic subunits (Figure 3.1) are present. Consequently, you will be able to note when one or more of the hybrids are missing from one of the strips.

The highest resolution of the isoenzymes of LDH on cellulose acetate is obtained by using a buffer made from Tris and barbital (Figure 3.7) (Monthony et al., 1978). Since this buffer, however, contains barbital, which is somewhat hazardous, an alternative (Tris-tricine), in which N-(2-hydroxy-1,1-bis(hydroxymethyl)ethyl)glycine (tricine) is used in place of barbital, is also described.

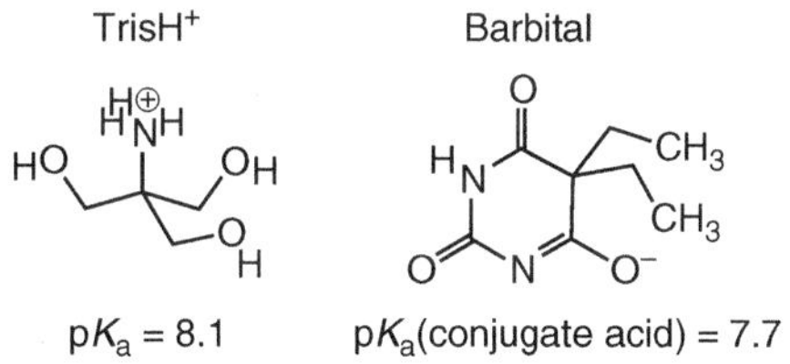

TrisH$^+$ Barbital

$pK_a = 8.1$ pK_a(conjugate acid) = 7.7

FIGURE 3.7 **Components of the electrophoretic buffer used for separation of the isoenzymes of LDH.** The conjugate acid (TrisH$^+$) of 2-amino 2-(hydroxymethyl)propane-1,3-diol (tris-hydroxymethylaminomethane; Tris) and the conjugate base of 5,5-diethylbarbituric acid (barbital). In the neutral conjugate acid of barbital, the proton is on the nitrogen, which is more basic than the oxygen, but oxygen bears most of the charge in the conjugate base because it is more electronegative.

Reagents, Supplies, and Equipment

Samples of protein to be assayed.
The supernates of the extracts of heart, kidney, liver, and skeletal muscle that you as a class prepared in the first exercise in this section will be characterized. If you have already purified the enzyme from the heart before you do this exercise, you should also submit both the original supernate and the purified enzyme to electrophoresis.

Commercial samples of LDH purified from skeletal muscle and heart serving as standards.
The dominant isoenzymes in these standards should be the homotetramers A_4 in skeletal muscle and B_4 in heart, but other isoenzymatic hybrids may be present in each. These act as standards that define the two extreme ends of the distribution.

Buffered solution for electrophoresis (two options):
Tris-barbital provides the highest resolution and is preferred (dispose after use in proper chemical waste):

61 mM barbital.
49 mM Tris.
Sodium ion from the sodium hydroxide used to raise the pH to 8.8.

or

Tris-tricine:

162 mM Tris.
48 mM tricine.
Chloride ion from the hydrochloric acid used to lower the pH to 8.6.

Components in the stain for activity of LDH:
  3 M lactate.
  75 mM NAD$^+$.
  10 mM 5-methylphenazinium methyl sulfate.
  3.7 mM Nitro Blue Tetrazolium chloride.

Polymeric matrix for electrophoresis and staining:
  Six strips of cellulose acetate (2.5 × 15.2 cm).

  Blotting strips and staining strips cut from Whatman filter paper.

Equipment:
  One electrophoretic apparatus that holds up to six strips of cellulose acetate: Semi-Micro II Chamber from Apacor (Figures 3.9 and 3.10) or equivalent.

  Plastic pan for hydrating strips of cellulose acetate and a small plastic tray for staining.

  P2 or P10 pipette and tips (more accurate than P20 for volumes less than 2 μL).

Precautions

Barbital buffer is toxic (details in Safety Guidelines for Biochemical Laboratories): wear gloves, and dispose of used buffer in hazardous waste.

To prevent the possibility of a deadly electrical shock, do not connect the leads from the power supply to the electrophoretic pan until the cover has been placed on the pan. Wait at least 20 s after the power supply has been turned off and disconnected before removing the lid and samples to be certain that all capacitance currents have dissipated.

Do not touch the strips of cellulose acetate with ungloved fingers as your fingerprints will create smears when the strip is developed. Use forceps and wear gloves.

Each strip of cellulose acetate must be completely wet before being submitted to electrophoresis to avoid funneling the current through a narrow region and burning the strip.

Procedure (One Laboratory Session)

Note regarding temperatures.
  Electrophoresis is performed in the cold room to prevent overheating of the strip caused by the heat of the electric current passing through the resistance of the cellulose acetate. Overheating would denature the enzyme. Staining, however, is done at room temperature because the activity of LDH is greater at this temperature than it is in the cold.

Assaying enzymatic activity in the samples and choosing samples to be submitted to electrophoresis (30 min).

- Inform your teaching assistant of the concentration of enzymatic activity for LDH in micromoles minute^{-1} milliliter^{-1} in the extract of heart, kidney, liver, or skeletal muscle that you prepared and assayed in LDH Exercises 1 and 2.

  Assaying of the standards.

- Assay the sample of commercial LDH from either skeletal muscle or heart that you have been assigned. The concentrations of enzymatic activity on the supplier's label were determined under different conditions than you will be using and are usually higher than you will observe. Calculating from the supplier's values, your teaching assistant will dilute each of these standards to a nominal concentration of enzymatic activity of 100 μmol min^{-1} mL^{-1}. You will receive 0.2 mL of a diluted standard. While one partner is preparing the strips for the electrophoresis, the other will perform an enzymatic assay on 50 μL of this sample. If the change in absorbance with time is too fast or too slow for an accurate assay, reassay a different volume or a different dilution of the sample.

- Calculate the concentration of enzymatic activity in this standard at the dilution that you were given, and report to the teaching assistant. The teaching assistant will write on the blackboard the numbers you and the other students report and calculate an average for the skeletal muscle and heart standards. The teaching assistant will then dilute a fresh sample from each of these two standard stocks to a concentration of enzymatic activity of 15 μmol min^{-1} mL^{-1}.

Preparing the samples from extracts of heart, kidney, liver, and skeletal muscle.

- Remove from the freezer a 2 mL aliquot of the extract you prepared in LDH Exercise 1, and allow it to thaw on ice.

While you are assaying the standards, the teaching assistant will prepare four mixtures to be used by all of the students in the section: one for heart, one for kidney, one for liver, and one for skeletal muscle. The teaching assistant will add a 2 mL aliquot of the extract from each group to the respective mixture, and will then calculate the final concentration of enzymatic activity (μmol min^{-1} mL^{-1}) of the mixture by determining the average of the activities of the mixed aliquots. It is best that none of the samples loaded on the electrophoresis strips have an activity greater than 30 μmol min^{-1} mL^{-1}, so any mixture of the extract from one of the tissues with a higher activity will be diluted.

Preparing the samples from LDH purified from bovine heart by affinity adsorption and affinity elution if you have already prepared this enzyme.

- Remove from the freezer one aliquot of the initial supernate and one aliquot of the purified enzyme. Allow them to thaw on ice.

- For these two samples, it is best if the concentration of enzymatic activity is no greater than 15 μmol min^{-1} mL^{-1}. If necessary, prepare an appropriate dilution or dilutions.

Preparing and conditioning the strips of cellulose acetate (15 min).

- While one partner is performing the enzymatic assay on the standard and preparing tissue samples for loading, the other, in the cold room, will hydrate the six strips in the buffered solution for the electrophoresis, mark them (Figure 3.8), and mount them in the apparatus (Figure 3.9).

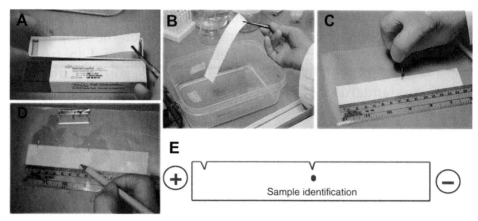

FIGURE 3.8 **Preparation of a strip of cellulose acetate for electrophoresis.** (A) The strip (2.5 × 15.2 cm) is handled with forceps or gloved fingers because the oil from your skin will interfere with the electrophoresis. (B) Before marking the strip, it is submerged in some electrophoretic buffer to hydrate it because dry strips are fragile. (C) The fully hydrated strip is placed on a piece of Parafilm. A razor is used to make two notches on one of the long edges. One notch is at the center of the strip to mark the origin where the protein will be deposited, and the other notch is made near one end to mark the end of the strip to be placed at the anode. (D) A #2 pencil is used at the opposite edge from the notches to write on the strip the sample that will be loaded. (E) The strip is ready for loading. The sample of protein will be applied to the origin at the center of the strip. Unlike electrophoresis on a gel of polyacrylamide, the proteins may migrate in either direction depending on their charge at the pH of the electrophoresis, so they are placed at the center of the field rather than at one end.

FIGURE 3.9 **Placement of strips of cellulose acetate and loading of sample.** The apparatus shown in this and subsequent figures is from Apacor (*Semi-Micro II Chamber*, item 51214). A similar device, however, from another supplier will work just as well. Everything is done in the cold room. (A and B) Buffered solution for the electrophoresis is poured into the reservoirs. The solution should cover the electrodes but not cross the divider between reservoirs for the anode and cathode. The levels of the solution in the two reservoirs should be as close to the same as possible to avoid siphoning of the fluid through the strips of cellulose acetate. The three strip holders, each in the shape of a bridge, are set into the tray. Refer to Figure 3.10 regarding the orientation of the anode and the cathode. If the red and black cords are connected to the power supply properly, and you are facing the end of the chamber with the electrical terminals, the anode will be on the left when the cover is placed on the tray. Each of the three strip holders has + and − symbols on it. The holders are placed in the tray in the appropriate orientation. Two strips are set in each holder with the notches marking the anode in the reservoir at the anode. Only the ends of the strips should be submerged. After all six strips are set in the holders, the cover of the apparatus is placed on the tray and the system is tested for a proper electrical current at the specified voltage as described in Figure 3.10. (C) The power supply is put in Standby mode, the cover of the apparatus is removed, and the correct sample of protein is loaded in the center of each strip.

- Allow the strips to soak for at least 10 min in the cold room. If any portion of the strip does not hydrate, which you can ascertain by looking for white blotches, discard the strip.

- Do not allow the strips to dry! If you are not ready to load them into the electrophoresis tray immediately after you have marked them, keep them under the buffered solution in the plastic tray.

Prepare the apparatus for electrophoresis (15 min).

- Pour the cold buffered solution for the electrophoresis, but not the soaked strips, from the pan in which you are soaking the strips into the reservoirs of the electrophoresis tray (Figure 3.9).

- Put electrophoretic buffer on each side of the electrophoretic chamber (note the precaution concerning barbital). Check to be certain that the buffer remains below the center divider along its entire length or the current will be short circuited between the reservoirs. Make sure that the level of the buffered solution is the same in the two reservoirs.

- Pick up a strip and allow excess buffer to drip off.

- Place it in a holder and orient it in the tray for the electrophoresis so that the notch marking the anode is on the anodic side of the tray.

- Submerge the end of the strip in the buffer in the appropriate reservoir to provide a complete circuit. The strip is submerged near its ends but not in the middle where the samples of protein will be loaded.

- Repeat with the other five strips.

- Close the tray for the electrophoresis, connect to the power supply, and test for current (Figure 3.10). Be certain the red (+) and black (−) cords are in the proper ports on the power

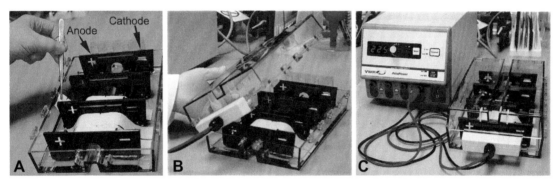

FIGURE 3.10 **Initiation of the electrophoresis.** (A) In addition to the markings on the strip holders, the cover on the tray for the electrophoresis is marked with + and − to indicate the orientation of the anode and cathode. (B) The cover for the apparatus is placed on the tray in the appropriate orientation and secured so that the cords are in contact with the terminals. (C) The power supply is switched back to Run mode to initiate the electrophoresis and the voltage is adjusted appropriately.

supply. For each strip, the notch at one end of the strip should be in the anodic reservoir. An equal amount of current should pass through each strip. This arrangement is parallel rather than series, so the total current shown on the display of the power supply should equal the sum of the currents through all the strips.

- Set the power supply for a constant voltage of 225 V (Tris-barbital) or 200 V (Tris-tricine). With Tris-barbital, expect about 1.2 mA for each strip, with Tris-tricine about 0.7 mA for each strip. A reading much higher than this indicates that there is a short circuit in the apparatus. A reading much lower indicates that one or more of the strips are not carrying current.

- After you are convinced that each strip is carrying its full load of current and that there are no shorts in the system, switch the power supply to Standby until you are ready to load.

Load the samples of protein (10 min).

- Before loading the samples, wait until excess liquid has evaporated from, or been adsorbed into, the strips. There should be no puddles of buffer on the strips.

- Use a P2 or P10 pipette to apply a sample to the center of each strip (Figure 3.9) according to the appropriate column in Table 3.1. Depending on the decision of the instructor, you will be running only six of these nine possible samples.

TABLE 3.1   Samples of LDH to be Submitted to Electrophoresis

	Extracts of Tissue	Purification of LDH from Heart
Standard for the A isoenzyme	1 μL	
Standard for the B isoenzyme	1 μL	
Extract of bovine heart	1 μL	
Extract of bovine kidney	1 μL	
Extract of bovine liver	1 μL	
Extract of bovine skeletal muscle	1 μL	
Initial supernate from bovine heart		1 μL
Purified LDH from bovine heart		1 μL
Purified LDH from bovine heart		2 μL

- Let the samples soak into the strips of cellulose acetate before proceeding.

Electrophoretic separation of the isoenzymes (2.5 h).

- When all of the samples have been loaded onto their individual strips, cover the pan (Figure 3.10; note precautions concerning electrical shock).

- Run the electrophoresis for 2.5 h at 225 V for the buffered solution of Tris and barbital or 200 V for the buffered solution of Tris and tricine. Record the current (mA) at the start and the end of the run. The separation of the molecules of protein by electrophoresis is only a function of the voltage, which determines the electrostatic force on the charged molecules. The current is carried by the ions in the buffered solution. The amperage registered on the power supply records current passing through the buffered solution. The current is carried mainly by the electrolytes surrounding the molecules of protein and is irrelevant to the migration of the molecules of protein themselves, but the current does produce heat.

- Turn off power supply, and wait for current to dissipate.

- While one partner removes the strips and marks them, the other partner should prepare the mixture for the stain from the stock solutions (Table 3.2), so that the stain is ready when the strips are ready to be stained. The final volume of mixture should be 11 mL.

- Remove the strip holder from the pan and the strips from the strip holders.

- Make two pencil marks on each strip, one at each of the points where the strip went below the surface of the buffer in the reservoirs. These points are easy to locate because the strip will be wet where it was below the surface of the buffer. Remember to handle the strips with gloves.

TABLE 3.2   Solution for Staining of the Activity of LDH

Component	Volume (mL)	Final Concentration (mM)
Lactate	1.4	380
NAD⁺	2.0	14
5-Methylphenazinium	0.7	0.63
Nitro Blue Tetrazolium	7.0	2.3

- After making these marks, blot the moisture from both of the ends that have been submerged in the buffer.

- Place the strips of cellulose acetate, protein side up, on paper towels.

Staining the strips of cellulose acetate (at room temperature) (20 min).

- Obtain pieces of Whatman filter paper of approximately the same length but twice as wide as the strips of cellulose acetate.

- Pour the prepared mixture of dyes and reactants into the small tray for staining.

- Submerge the strips of filter paper in the dye–substrate mix so that they absorb the stain.

- Lay a soaked strip of filter paper over each strip of cellulose acetate making sure there are no air spaces preventing contact (one soaked filter can usually cover two strips of cellulose acetate).

- Leave the filter paper saturated with the staining solution in contact with the cellulose acetate at room temperature until distinct blue spots develop on the nitrocellulose. Intense spots may be visible through the Whatman paper, but for most spots it will be necessary to turn over the strips of cellulose acetate and filter paper once in a while to watch the spots develop. The spots should appear in 5–10 min.

Preparing a photographic image and recording the distances migrated by each of the hybrids (10 min).

- Lay the stained strips of cellulose acetate on a sheet of white printer paper. As you do, space the six strips on the field so that you can capture all of them in a single image. This will simplify analysis of the data. Be certain the strips of cellulose acetate are all in the same orientation with the anodic end on the left, but align the notches that marked the origins on each strip where the samples were applied.

- Place a clear, plastic ruler (metric) along the outside edge of the array. Align one of the centimeter marks in the middle of the ruler with the position of the origins.

- Record a photographic image of the strips while they are still wet. This image of the stained strips is recorded with reflected light. Place the light source above the strips rather than below as you did for the stained gel of polyacrylamide, which, unlike the cellulose acetate, was transparent. If you are using the Kodak Molecular Imaging (MI) software, use the White Light, Epilumination setting. After capturing an image, convert it to JPG format because the BIP format can only be opened with Kodak MI. Be certain you know which sample was on which strip and the orientation of cathode and anode. You should draw a diagram of the image in your notebook and label the diagram with this information.

- Measure, in millimeters, the distance that the spots moved from the origin and enter that information into your notebook. Be sure to indicate whether spots traveled towards the cathode (+) or the anode (−). Also, for each strip, record the distance between the two pencil lines marking where the strip emerged in turn from the two reservoirs.

Analysis of the Data and Questions to Address in Your Report

1. Calculate the electrophoretic mobility of each component. The electrophoretic mobility $(mm^2\,h^{-1}\,V^{-1})$ is the rate at which the protein moved $(mm\,h^{-1})$ divided by the magnitude of the electric field $(V\,mm^{-1})$. The total drop in voltage across the strip from the point at which it emerges from the reservoir at the anode to the point at which it submerges into the reservoir at the cathode is equal to the voltage reading on the power supply because the resistance within the reservoirs is negligible. The magnitude of the electric field is the voltage registered on the power supply (V) divided by the distance between these two points (mm). You have already marked these two points with lines. The sign on the electrophoretic mobility indicates the charge on the protein; migration towards the cathode is positive, and migration towards the anode is negative.

2. Discuss the relation between the electrophoretic mobility (both sign and magnitude), the net charge on the protein, and the resistance of the medium that is sieving the proteins.

3. There are five possible hybrids of the isoenzymatic A and B subunits (Figure 3.1). Use the results for the standards from skeletal muscle and heart as a guide, and assign the subunit composition of each protein spot on the strips for the four extracts from the tissues.

4. If you did not know that LDH was a tetramer, how would the patterns of these hybrids tell you that it is a tetramer? What else could it be, based solely on these patterns?

5. It is not possible to assign exact values of p*I* to the hybrids of LDH with an electrophoresis at only one pH. Relative values, however, may be determined. Rank the various hybrids in order of increasing p*I*.

6. How would you use electrophoreses at several different values of pH but at the same concentration of electrolyte to estimate the actual values of p*I* for each of the hybrids?

7. In mammalian cells, the interconversion of pyruvate and lactate may be catalyzed in either direction by LDH. The physiological circumstances that determine the net direction of the reaction were discussed earlier in this section. Your data should allow you to make some conclusions regarding how much of each isoenzyme is expressed in each of the tissues examined. Remember that in each isoenzymatic hybrid, the A subunits and the B subunits, which are the fundamental units of the isoenzymes of LDH, operate independently of each other, so it is only the distribution of A and B subunits among the tissues that is relevant. Discuss any correlation you see between expression of the A and B subunits and the predominant direction of catalysis of the reaction in these tissues. Such a correlation has been reported in the literature (Berg et al., 2012; Saad et al., 2006; Voet and Voet, 2010).

LDH EXERCISE 4

Fractionation of Proteins in Bovine Heart by Precipitation with Sulfate

Technical Background and Experimental Plan

Tissue from bovine heart will be homogenized, and proteins will be selectively precipitated out of solution using step-wise increases in the concentration of ammonium sulfate (Figure 3.11). The high concentration of sulfate ion, which is an ion that is unable to enter the shell of hydration around the protein (Figure 3.12), causes the bulk solution to become thermodynamically

Weigh out 80 g of chopped bovine heart.

Fragment the tissue in the blender with 320 mL of 50 mM buffered solution (Figure 1.7).

Homogenize the blended tissue in a homogenizer with a teflon pestle (Figure 3.4)

Transfer the homogenate into two 250 mL bottles for the centrifuge.

Centrifuge the homogenate at 13,000 rpm (26,000 × g) for 20 min at 4 °C.

Discard the pellet.   Remove ten aliquots of 0.5 mL from this initial supernate.

Weigh out an amount of solid ammonium sulfate equal to 0.226 g times the number of milliliters in the supernate.

Add this ammonium sulfate slowly over 15 min to produce 39% saturation.
Equilibrate for 10 min.

Centrifuge the suspension as you did the homogenate.

Redissolve the 39% pellet in 20 mL of 50 mM buffer and remove four aliquots of 0.5 mL from the redissolved 39% pellet.   Remove four aliquots of 0.5 mL from the 39% supernate.

Weigh out an amount of solid ammonium sulfate equal to 0.187 g times the number of milliliters now in the supernate.

Add this ammonium sulfate slowly over 15 min to produce 68% saturation.
Equilibrate for 10 min.

(If you are now out of time, you may stop here and store everything at 4 °C until the next session. Proteins remain in their native state for decades as precipitates in ammonium sulfate at 4 °C .)

Centrifuge the suspension as you did the homogenate and the 39% precipitate.

Redissolve the 68% pellet in 20 mL of 50 mM buffer and remove four aliquots of 0.5 mL from the redissolved 68% pellet.   Remove four aliquots of 0.5 mL from the 68% supernate.

Save the remainder of the 68% supernate and the remainder of the redissolved 68% pellet at 4 °C.

Assay the initial supernate, 39% pellet, 39% supernate,
68% pellet, and 68% supernate for activity of LDH.

Dialyze the redissolved pellet that contains the majority of the activity of LDH.

FIGURE 3.11 **Flow chart for the fractionation of the supernate of the extract from bovine heart by precipitation with sulfate ion.** A 100% saturated solution of ammonium sulfate is a solution at equilibrium over an excess of solid ammonium sulfate. A saturated solution of ammonium sulfate at 4 °C is the solution produced by adding 71.7 g of ammonium sulfate to 100 mL of water, for which the final volume is 138 mL. This is, by definition, a 52% solution of ammonium sulfate (52 g in each 100 mL of solution), or a 3.9 M solution of ammonium sulfate. You should come to an understanding of the difference between percent saturation and percent.

A

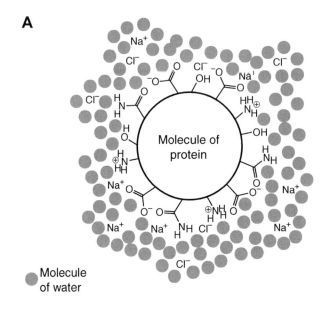

Molecule of water

B

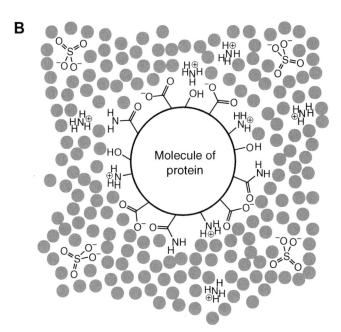

FIGURE 3.12 **Addition of inorganic ions to decrease the solubility of a protein through preferential solvation.**   Most of the surface of a native molecule of protein is covered with side chains from the amino acids in the folded polypeptides. All proteins that are soluble in aqueous solution have, distributed over most of their surface, ionized side chains that have donors and acceptors for hydrogen bonds and side chains that are not ionized but that also have donors and acceptors for hydrogen bonds. These proteins are soluble because the donors and acceptors for hydrogen bonds form hydrogen bonds with the molecules of water in the solution. These hydrogen bonds permit the molecules of protein to become incorporated into the network of hydrogen bonds that is liquid water. In addition to these hydrogen bonds, the charges on the surface of the protein are strongly hydrated and any hydrophobic surface has a layer of ordered water between it and the bulk solution. All of these interactions create a layer of hydration around the protein. The water in this layer is distinct and has different properties from the water in the bulk solution. (A) When a molecule of protein is dissolved in a solution of sodium chloride, neither the sodium ion nor the chloride ion shows any significant preference between dissolving in the water in the bulk solution and dissolving in the layer of hydration, and these two ions distribute equally between these two phases. (B) When a molecule of protein is dissolved in a solution of ammonium sulfate, the sulfate has a strong preference for being dissolved in the water in the bulk solution and is excluded from the layer of hydration around the protein (Arakawa and Timasheff, 1984). The ammonium ion has a moderate preference for the bulk solution and is modestly excluded from the layer of hydration around the protein (Cacace et al., 1997). Nevertheless, in both the case of sodium chloride and the case of ammonium sulfate, the solution immediately surrounding the protein must contain an excess of cations, sodium ions or ammonium ions, or an excess of anions, chloride ions or sulfate ions, equivalent to the net charge of the protein so that the solution in this local region is electroneutral.

incompatible with the layer of hydration around the protein and promotes the precipitation of the protein (Arakawa and Timasheff, 1984). Each molecule of protein, depending on its concentration in the solution, the distribution of the side chains of the amino acids on its surface, and the peculiarities of its shell of hydration precipitates in a different range for the concentration of sulfate ion. Precipitated proteins that have been *salted out* of the solution are separated from those that remain dissolved by centrifugation after each increase in the concentration of sulfate ion (Berg et al., 2012; Boyer, 2012; Kyte, 2007, pp 21–23; Lehninger et al., 2008; Ninfa et al., 2010; Voet and Voet, 2010).

The resulting fractions will be assayed for the activity of LDH. The fractions containing most of the activity of LDH will be pooled, and the pool will be dialyzed to remove the ammonium sulfate and other small molecules (Figure 3.13). Aliquots of the dialyzed pool will be stored for use in later experiments.

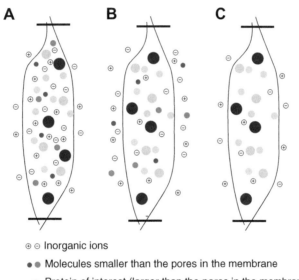

⊕ ⊖ Inorganic ions

● ● Molecules smaller than the pores in the membrane

⬤ Protein of interest (larger than the pores in the membrane)

⬤ ● Other proteins larger than the pores in the membrane

FIGURE 3.13 **Dialysis to eliminate inorganic ions and small molecules from a solution of protein.** The membrane for dialysis is a porous, flexible, transparent plastic typically supplied in the form of tubing. A bag for dialysis is formed by cutting an appropriate length of the tubing, using a clip to seal one end, filling the resulting bag with the solution of protein, and sealing the other end with a second clip (Figure 3.14). The sealed bag is submerged in a solution in a beaker or flask. This solution has the pH and ionic composition chosen by the investigator to become the pH and ionic composition of the solution of protein. The salts and small molecules in the solution outside the bag replace the salts and small molecules inside the bag because they all can easily pass through the pores in the plastic. In the end, the protein is dissolved in the solution against which it was dialyzed. The size of the pores in the membrane determines the size of the proteins that are retained in the bag because they cannot pass through those pores. The size of the proteins retained by a particular type of dialysis tubing, for historical reasons, is reported as a molar mass. (A) The bag contains proteins of a wide range of sizes and a high concentration of ammonium sulfate. (B) As diffusion occurs across the membrane, the concentration of all molecules smaller in size than the pores in the membrane decreases inside the bag. (C) After a few replacements of the solution in the beaker, the concentration of electrolytes and small molecules and the pH of the solution in which the proteins are dissolved are the same as the solution against which it was dialyzed.

Reagents, Supplies, and Equipment

Bovine heart that has been cubed and passed through a meat grinder (in cold room).

Buffered solution for homogenization: 10 mM EDTA, 50 mM sodium phosphate, pH 7.5.

Buffered solution for dialysis: 10 mM EDTA, 20 mM sodium phosphate, pH 6.0.

Solid ammonium sulfate, analytical reagent or enzyme grade.

Dialysis tubing with a pore size appropriate for retaining molecules of protein with molar masses greater than 10,000 to 30,000 g mol^{-1} (Figure 3.14), stored in the dialysis buffer at 4 °C.

FIGURE 3.14   **Initiating the dialysis of a sample of protein.**   The membranes for dialysis are typically supplied dry. The tubing will have been rehydrated and in an appropriate buffer when you need to use it. Wear gloves and cut a segment of tubing of the appropriate length. A beaker of buffered solution for the dialysis is prepared. A plastic clip is used to seal one end of the tube. (A) The redissolved pellet is pipetted into the bag. (B) The sample is sealed in the bag by attaching a second plastic clip. (C) The sealed bag is allowed to float in the buffer long enough for complete equilibration of inner and outer solutions. The buffered solution in the beaker is usually replaced twice during this procedure.

Two centrifuge bottles with a volume of 250 mL (Figure 1.6) and screw-caps. The rubber O-ring in the cap is essential for its seal!

Kitchen blender (Figure 1.7, cold room).

Potter–Elvehjem homogenizer with Teflon pestle (motor-driven, Figure 3.4).

Beckman J2 preparative centrifuge and JA-14 rotor or equivalent.

Precautions

Keep all items that come into contact with the tissue and extracts, such as containers, buffers, and centrifuge tubes, on ice.

Do not use the centrifuge without supervision.

Balance the tubes as accurately as possible before loading them into a rotor for the centrifuge.

Before beginning a dialysis, check for leaks in the bag by first filling the bag with water and squeezing (Figure 3.14). When sealing the bag, do not put the clamps on the very end of the bag since they may fall off. Fold over the tubing, and then clamp the tubing well below the fold.

Procedures (Two Laboratory Sessions and an Overnight Step)

PART 1: HOMOGENIZE THE TISSUE, REMOVE DEBRIS, AND PRECIPITATE THE PROTEINS

Preparation of a homogenate of the tissue. The blending and homogenization are to be done in the cold room (1.5 h).

- Go to the cold room and weigh out around 80 g of the chopped heart. Use one of the triple beam balances. You do not have to have exactly 80 g, but record the actual mass.

- Mix in a beaker the 80 g of tissue with 4 mL (g of tissue)$^{-1}$ of cold 50 mM sodium phosphate, 10 mM EDTA pH 7.5, and bring the suspension to less than 5 °C on ice.

- Fragment the ground tissue in a blender (Figure 1.6). Blend at the highest speed for a total of 3 min. Blend only 30 s at a time, and check the temperature between each 30 s period. Keep the temperature below 20 °C. After you have blended the tissue, pour the suspension back into the beaker on ice.

- Disrupt the cells in the suspension of fragmented tissue using a motor-driven Potter-Elvehjem homogenizer (Figure 3.4). Homogenize the 400 mL of blended tissue in 50 mL batches. Keep the temperature of the homogenate below 20 °C.

- Transfer the homogenate to two 250 mL centrifuge bottles (Figure 1.6). Balance the two bottles on a double pan balance by transferring homogenate back and forth. After balancing them, be certain the caps are tight. Any leakage will contaminate the centrifuge rotor, which you will have to clean.

- Centrifuge at 13,000 rpm (26,000 × g) for 20 min at 4 °C. While the centrifugation is running, clean the blender and the homogenizer, and set up the magnetic stir plate.

- Decant the initial supernate into a graduated cylinder, and record the volume.

- Discard the pellet that contains cellular debris.

- Save ten 0.5 mL aliquots of the initial supernate and label them. Add glycerol to 10% (two drops for each tube). Quick freeze these aliquots in a slurry of dry ice in isopropanol and store at −20 °C. Refer to LDH Exercise 1 for instructions on labeling and freezing the aliquots. These aliquots may be used for the following assays: enzymatic activity, concentration of protein (Bradford assay), separation of isoenzymes by electrophoresis, electrophoresis of the native enzyme on a gel of polyacrylamide, electrophoresis of the polypeptides unfolded with dodecyl sulfate on a gel of polyacrylamide, and immunoblotting.

First precipitation with ammonium sulfate (1.5 h).

- To the remaining supernate, add gradually, over a period of 15 min, 0.226 g of solid ammonium sulfate for every milliliter of the supernate while constantly stirring the solution with a magnetic stir bar and a magnetic stir plate in the cold room. Record the amount of ammonium sulfate you added.

- Stir slowly an additional 10 min to dissolve all of the ammonium sulfate and equilibrate, then transfer to two centrifuge bottles.

- Centrifuge the cloudy suspension in the two bottles as you did with the homogenate.

- Do not discard any of the fractions you are collecting until you have determined which fraction has the most enzymatic activity; you will use that fraction in subsequent exercises. Keep all of the samples you have collected and will collect on ice or in the cold room.

- Transfer the supernate from this centrifugation to a fresh container. Determine the total volume (sum of both centrifugation bottles).

- Remove four 0.5 mL aliquots. Label aliquots 39% supernate.

- Save the remaining 39% supernate on ice.

- Redissolve the two pellets together in a total of 20 mL cold 10 mM EDTA, 50 mM sodium phosphate, pH 7.5.

- Remove four 0.5 mL aliquots, and label the tubes 39% pellet. Label the remainder of the redissolved 39% pellet.

- Add glycerol to 10% to each 0.5 mL aliquot and quick-freeze them as described above. Store these aliquots at −20 °C. Store the remainder of the redissolved 39% pellet at 4 °C. The 0.5 mL aliquots of the 39% supernate and 39% pellet will be used for assays of enzymatic activity and concentration of protein (Bradford) assays.

Second precipitation with ammonium sulfate (15 min in this session followed by overnight storage).

- To the remaining 39% supernate, add gradually over 15 min 0.187 g of solid ammonium sulfate for every milliliter of supernate, as above.

- If you are out of time, you may stop here and store the slurry at 4 °C for the next period. Enzymes are unusually stable as precipitates in ammonium sulfate at 4 °C. Do not freeze this sample!

PART 2: SPIN DOWN THE 68% PRECIPITATE, ASSAY THE ENZYMATIC ACTIVITY IN THE FRACTIONS, AND BEGIN THE DIALYSIS

Spin down precipitated proteins (1 h).

- Transfer the slurry in 68% ammonium sulfate to two centrifuge bottles, and centrifuge as you did for the homogenate and the 39% precipitate.

- Record the volume of this third supernate, and then remove four 0.5 mL aliquots. Label the tubes 68% supernate. Store the remainder of the 68% supernate on ice or at 4 °C.

- Redissolve the two pellets of 68% precipitate in a total of 15 mL 50 mM buffer as above and remove four 0.5 mL aliquots. Label these tubes 68% pellet. Store the remainder of the redissolved 68% pellet at 4 °C.

- Add glycerol to 10% to each 0.5 mL aliquot of the 68% supernate and redissolved 68% pellet and freeze as described in LDH Exercise 1. These 0.5 mL aliquots will be used for assays of enzymatic activity and concentration of protein (Bradford) assays.

Identifying the precipitate with the majority of the activity of LDH (1.5 h in this session followed by an overnight step).

At this point you should have three major, unfrozen fractions that have seen ammonium sulfate: the 39% redissolved pellet, the 68% redissolved pellet, and the 68% supernate. Now you must determine which of the three contains the majority of the enzyme. The 0.5 mL aliquots have been frozen but the majorities of each of these fractions have been stored at 4 °C. Remove samples from these three solutions for the assays of enzymatic activity.

- Dilute samples of the 39% pellet, 68% pellet, and 68% supernate appropriately with 100 mM sodium phosphate, pH 7.5. Assay for activity in triplicate, as described in LDH Exercise 2.

- Calculate the total activity for each of these three fractions as described in LDH Exercise 2. Convert the dA_{339}/dt to the units of micromoles minute^{-1} milliliter^{-1}, and multiply the concentration of enzymatic activity by the total volume of each of these fractions to obtain the total activity of LDH.

- If the majority of the activity is in either the redissolved 39% pellet or the redissolved 68% pellet, dialyze that fraction against 1 L of 10 mM EDTA, 20 mM sodium phosphate, pH 6.0.

- If most of the activity remains in the 68% supernate, you will have to continue with a further precipitation at a concentration of 85% ammonium sulfate and again check the enzymatic activity of both supernate and redissolved pellet before proceeding with the dialysis. If you need to do the 85% precipitation, add another 0.115 g of ammonium sulfate for each milliliter of the supernate. Mix, equilibrate, and centrifuge. Remove and freeze aliquots of the 85% fractions and assay as for the other fractions.

- Dialysis (Figure 3.14) is performed at 4 °C for at least 40 h in a beaker. Place a stir bar in the beaker, and gently stir the bag and its surrounding solution in the cold room. Remember to make sure that the stir plate has warmed up and is stirring gently before you leave. The dialysis buffer will need to be changed two times before the next laboratory session. You or your partner must come in tomorrow to change the buffered solution in the beaker in the morning and in the evening. The dialysate will be used in the next exercise.

Assay activity of LDH in aliquots of the initial supernate and the 39% supernate (30 min).

- Thaw one aliquot each of the initial supernate and the 39% supernate. Assay for enzymatic activity of LDH, and calculate the concentration of activity for LDH and the total activity in the initial supernate and the 39% supernate. This information will be needed to complete the table summarizing the purification in a later exercise.

- Check to make sure that the total activities you just measured are consistent with total activities for the 39% pellet, the 68% pellet, and the 68% supernate. For example, the sum of the total activities in the 39% pellet and the 39% supernate should equal the total activity in the initial supernate, and so forth. Correct the total activities for the volumes of the aliquots you removed before making these comparisons.

Analysis of the Data and Questions to Address in Your Report

Questions 1 and 2 were also asked at the end of LDH Exercise 1. They are repeated here in case that exercise is not on the schedule for the course.

1. Refer to FNR Exercise 1 and LuxG Exercise 1. Even if you have not performed these exercises, the sections in these exercises covering technical background, experimental plan, and procedures should help you to answer this question and the next. Consider the different methods used to lyse cells in the leaves of spinach, cells of *E. coli*, and cells in mammalian tissue to release their cytoplasm. Discuss the structure of the tissues and cells and explain the reasons for the variation in technique.

2. Consider the centrifugations used to pellet chloroplasts and cells of *E. coli*. The homogenization of mammalian cells with a Potter–Elvehjem homogenizer does not disrupt nuclei or mitochondria (Boyer, 2000; de Duve et al., 1955). Discuss why the procedures specified in FNR Exercise 1 and LuxG Exercise 1 suggest that the mammalian subcellular organelles were pelleted during the centrifugation in the current exercise.

3. In FNR Exercise 3 and LuxG Exercise 6, ultrafiltration under centrifugal force was used to decrease the concentration of ions and small molecules in the preparations of the respective enzyme. Discuss whether it would have been practical to use this technique rather than dialysis in the current exercise.

4. You have identified the fraction from the homogenate of the heart with the highest concentration of activity. Discuss whether or not it is likely that the enzyme is pure at this point.

5. Any salt of sulfate will cause proteins to precipitate from solution if it can be made concentrated enough. There are several reasons that ammonium sulfate is used almost exclusively for this purpose. What do you think these reasons are?

6. Describe how ammonium sulfate could be used to concentrate a solution of protein.

LDH EXERCISE 5

Purification of LDH from Heart by Affinity Adsorption and Elution

Technical Background

Purification by affinity is the isolation of a protein based on its interaction with a ligand for which it has a high affinity (Berg et al., 2012; Boyer, 2012; Kyte, 2007, pp 26–30; Lehninger et al., 2008; Ninfa et al., 2010; Voet and Voet, 2010). A ligand is any molecule that binds specifically to a site on the surface of a protein. Ligands can be substrates for an enzyme, natural molecules the protein has evolved to bind, or synthetic molecules that are analogues of these natural ligands. The dissociation constant of the ligand from the site on the protein to which it binds is a quantitative measure of the affinity of the protein for the ligand. The units of the dissociation constant are molarity. A useful fact to remember is that when the molar concentration of the protein is significantly less than the dissociation constant for one of its ligands, the molar concentration of the ligand free in solution needed to occupy, at equilibrium, half of the sites on the molecules of protein is equal to the dissociation constant of the ligand from the protein. The smaller the dissociation constant, the tighter the binding of the ligand to the enzyme or protein and the higher its affinity for the site. Ligands with dissociation constants in the micromolar range are tight ligands, but there are some ligands the dissociation constants for which are in the nanomolar range. In the experiments described here, the ligand bound to the solid phase, adenosine-5′-monophosphate (AMP), has a dissociation constant of 5 mM from LDH of *Squalus acanthias* (Mosbach et al., 1972) and the ligand used to elute LDH from the solid phase has a dissociation constant of 20 μM from B_4 LDH of *Gallus gallus* (Everse, 1971).

Purification by affinity has two equally important aspects. The solid phase to which a ligand for the protein of interest is attached can be a solid phase like the solid phase of Procion-Red-cellulose or Cibacron-Blue-agarose that was used to purify FNR in FNR Exercise 3. To make such an affinity adsorbent, a ligand that has a significant affinity for the protein being purified is attached covalently to a solid phase. The protein is adsorbed to the solid phase for which it has a particular affinity, and after other proteins that do not have an affinity for the ligand on the solid phase have been washed away, the adsorbed protein is eluted. In the case of FNR, the protein is eluted nonspecifically with increasing concentrations of potassium chloride, and no advantage is taken of affinity during the elution. It is possible, however, to use a ligand for which it also has an affinity to elute the protein. In fact, affinity elution is even more reliable than affinity adsorption because the ligand that is used to elute the protein is not attached to a solid phase so it is more accessible to the sites on the protein. It is possible to adsorb a protein to a solid phase for which most proteins have an affinity, and then elute only the protein of interest by using a ligand for which only that protein has a significant affinity.

The sequence of events leading to the dissociation of an enzyme such as LDH from the solid phase during affinity elution is as follows. The ligand attached covalently to the solid phase associates with the active site of the enzyme in a reversible equilibrium with a particular dissociation constant. Because the dissociation is at equilibrium, the enzyme is constantly dissociating and reassociating with ligands attached to the solid phase at all times. These dissociations and reassociations are governed by specific rate constants. When there are no other ligands for the enzyme present, the enzyme spends most, but not all, of its time associated with the covalently attached ligands and moves very slowly through the solid phase, advancing only during the short intervals when it is not associated. If another ligand with a higher affinity for the active site is introduced into the solution surrounding the solid phase, it will associate with the active site during the short period that the enzyme has dissociated from the covalently attached ligand and before it can associate with another covalently bound ligand. Once the other ligand, which is free in solution, has associated with the active site, the active site is occupied most of the time with this ligand in solution and can rarely reassociate with the covalently attached ligands, and the enzyme now spends most of its time free in solution rather than being bound to the solid phase. It now washes through the column rapidly.

In the present exercise, LDH is adsorbed to a solid phase to which AMP is attached (Figure 3.15A). The affinity of LDH for AMP free in solution is not very strong. It has a dissociation constant of around 5 mM, and attaching the AMP to a solid phase through its adenine certainly makes it

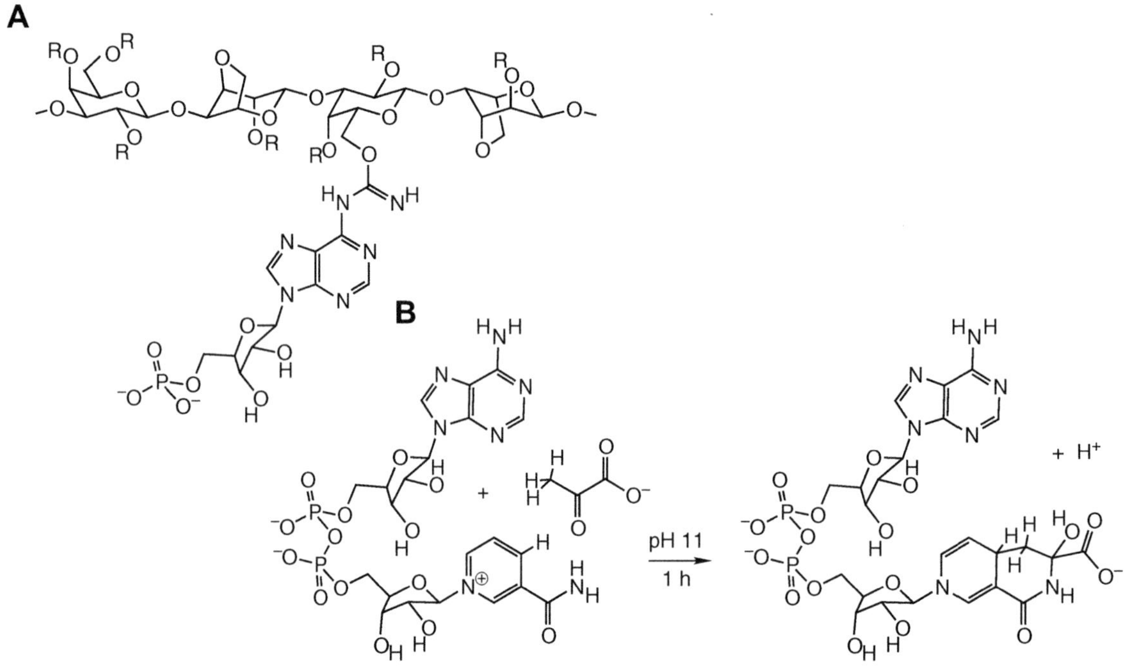

FIGURE 3.15 **The ligands used for purification of LDH by affinity adsorption.** (A) Adenosine-5′-monophosphate covalently linked to agarose (Cuatrecasas, 1970a, b), which is provided commercially in beaded form. It is perhaps because it corresponds to a part of NAD that AMP is an effective ligand for LDH when it is covalently attached to the solid phase, even though it would seem that the adeninyl group would be sterically inaccessible to its active site. (B) The adduct between pyruvate and NAD⁺. The adduct is synthesized by mixing NAD⁺ with pyruvate and adding base to promote the deprotonation of the methyl carbon of pyruvate and form its enolate. The carbanionic carbon of the enolate adds to the electrophilic position at carbon 4 on the nicotinamide, the position to which a hydride ion is transferred, and then the conjugate base of the amido group on the nicotinamide adds to the carbonyl of the pyruvate to close the second ring. Of the dehydrogenases tested, this adduct is specific for LDH, and it has a high affinity for the active site of the enzyme, binding with a dissociation constant of around 20 μM (Everse, 1971). It is this adduct that provides the specificity of the procedure for purifying LDH.

an even weaker ligand. Nevertheless, LDH does adsorb to the solid phase, probably because the density of the ligands is so high. Other proteins, however, also have affinity for AMP and also adsorb to this solid phase (Mosbach et al., 1972), and they cannot be separated by affinity adsorption alone from LDH. Consequently, it is the affinity elution that is used in this exercise that actually purifies LDH. The ligand chosen for affinity elution is an adduct between NAD⁺ and pyruvate (Figure 3.15B). Of the dehydrogenases tested, this adduct was specific for only LDH, and it has a reasonably high affinity for the active site of the enzyme with a dissociation constant of around 20 μM (Everse, 1971). Under the conditions used, only LDH binds this ligand in preference to the AMP attached to the solid phase, and the LDH is prevented from reassociating with the solid phase when it is bound to this ligand because the ligand also occupies the site on LDH for AMP. Once it can no longer bind to the solid phase, LDH can be washed out of the column as the pure enzyme.

Experimental Plan

A column containing a matrix of beaded agarose to which AMP has been covalently bound is prepared (Cuatrecasas, 1970a, b). The redissolved and dialyzed ammonium sulfate precipitate, containing the majority of the enzymatic activity, is applied to this column (Figure 3.16). The

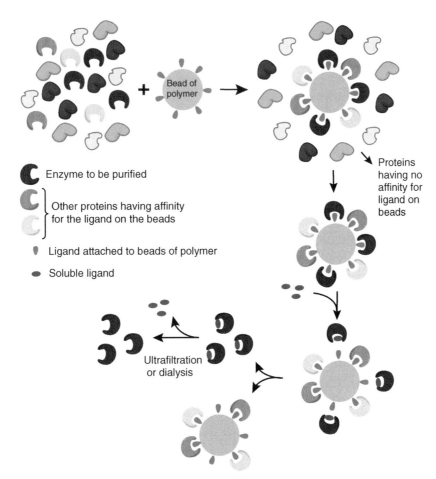

FIGURE 3.16 **Purification of an enzyme by its affinity for particular ligands.**   A sample containing a mixture of proteins is applied to a column of beaded solid phase (gray circles) to which an appropriate ligand for the active site of the enzyme (green ovals) has been covalently attached. Only the enzymes with specificity for the attached ligand bind, and unbound proteins are washed through the column. A soluble ligand (red ellipses), different from the ligand attached to the solid phase, for which the active site of the enzyme of interest also has an affinity is then applied. The soluble ligand associates with the active site of the enzyme when it has temporarily dissociated from the ligand covalently attached to the solid phase, and this association prevents the enzyme from reassociating with the solid phase. The complex between this other ligand and the enzyme is washed out of the solid phase. The strategy of relying on the specificity of the enzyme for two different ligands eliminates all proteins that have no affinity for either and proteins that have an affinity for only one of the two ligands. If the ligands are carefully chosen, this two-step procedure should eliminate all other proteins but the enzyme of interest. The concentration of soluble ligand in the solution is lowered by ultrafiltration under centrifugal force or by dialysis. Lowering its concentration in solution causes the ligand to dissociate from the active site. Because ultrafiltration under centrifugal force is more rapid than dialysis, which is, however, more effective, the former procedure will be used in this exercise.

AMP–agarose selectivity binds LDH and proteins with an affinity for AMP while most of the other proteins pass through the column (Kaplan et al., 1974). The LDH is then specifically eluted from the AMP–agarose by adding the adduct of NAD$^+$ and pyruvate (Everse, 1971). The fractions containing the eluted LDH are then identified by assaying the fractions for the enzymatic activity of LDH. The fractions containing the most enzymatic activity are pooled and dialyzed for use in subsequent experiments.

Reagents, Supplies, and Equipment

Commercially available beaded AMP–agarose (~5 mL of slurry) in 10 mM EDTA, 20 mM sodium phosphate, pH 6.0. This expensive material should be recovered for reuse.

10 mM EDTA, 20 mM sodium phosphate, pH 6.0.

30 mL of 0.1 mM adduct of NAD^+ and pyruvate, 10 mM EDTA, 20 mM sodium phosphate, pH 6.0, prepared on the day it will be used.

10 mM EDTA, 50 mM sodium phosphate, pH 7.5.

Reusable opaque plastic Oak Ridge tubes with a volume of 38 mL, a thick (2–3 mm) wall, and a round bottom that are strong enough to withstand spins at high speed (13,000 rpm) (Figure 1.6).

Cheesecloth and funnel for filtering.

Amicon Ultra-4 30K ultrafiltration devices with a capacity of 4 mL. (Figure 1.16). The designation 30K is the abbreviation for the molar mass of the smallest spherical protein (30,000 g mol^{-1}) that is unable to pass through the pores in the membrane. Two or four are needed.

Plate reader (Appendix I and Figure 1.13), conventional spectrophotometer or Spec 20 spectrophotometer (Appendix II).

Beckman J-2 refrigerated preparative centrifuge and JA-14 rotor, for which 7000 rpm equals $7500 \times g$ and 13,000 rpm equals $26,000 \times g$, or equivalent.

Chromatographic apparatus and affinity adsorbent:

The apparatus is assembled as was the apparatus described in FNR Exercise 6 (Figures 1.24, 1.25, and 1.26) except that the column is not nearly so large.

Glass column with two-way stopcock at bottom, a diameter of 1 cm, and a height of 10 cm, as described in FNR Exercise 3 (Figure 1.17).

Three-way stopcock on the cap of the column with a 10 mL syringe attached to the side port.

Mariotte bottle with a stopper into which a glass tube is inserted.

Adapters and tubing.

Fraction collector.

Precautions

The AMP–agarose column should not run dry at any time. To prevent this from happening, the outlet of the tubing at the fraction collector should always be above the level of the top of the AMP–agarose in the column.

Do not use the centrifuge without supervision.

Procedures (Two Laboratory Sessions)

PART 1: AFFINITY ADSORPTION AND ELUTION

Preparation of the dialyzed ammonium-sulfate fraction of the extract from heart (1 h).

- Remove the ammonium sulfate fraction from the dialysis bag and transfer it to a 38 mL Oak Ridge tube on ice.

- Centrifuge the dialyzed ammonium sulfate fraction at 13,000 rpm for 15 min. Balance the tubes, one from each group, by adding purified water to one of the pair.

- Transfer the supernate to a fresh tube while avoiding any particles that might clog the column; for example, there may be some lipid floating on the surface. Discard any pellet.

- Filter the supernate through a minimum amount of cheesecloth, which will absorb some of your solution, to remove any residual particles. Record the volume of the filtered supernate, and put it on ice.

- Remove five 0.3 mL aliquots from the supernate and label them "precolumn LDH." Add one drop of glycerol to each portion, and freeze them in a slurry of dry ice in isopropanol. You will need to assay one of these aliquots for enzymatic activity at some point to complete the table summarizing the purification at the end of LDH Exercise 6. Other aliquots will be used for electrophoresis of the polypeptides unfolded with dodecyl sulfate and for immunoblotting.

- Keep the remainder of the supernate on ice while you prepare the column.

Preparation of the column for affinity adsorption and setting up the fraction collector (60 min).
    Refer to FNR Exercise 6 and Figures 1.24, 1.25, and 1.26.

- Attach the plastic column to the rack in the cold room.

- Close the stopcock, and fill the column halfway with 10 mM EDTA, 20 mM sodium phosphate, pH 6.0.

- Add a slurry of AMP–agarose and open the stopcock. Add slurry to the column while it is flowing. Layer on 10 mM EDTA, 20 mM sodium phosphate, pH 6.0, while the bed is settling so that it does not run dry. Do not let the column run dry! The final packed volume, the bed volume of the column, should be around 3 mL.

- Set up the fraction collector, and attach the outlet of the column to the tubing feeding the fraction collector. Learn how the fraction collector works before loading the dialyzed ammonium sulfate fraction or you may lose your enzyme. Set the Mode switch to Drop. Adjust the number of drops for each fraction with the up and down arrows when the Run/Stop switch is in the Stop position. Usually, 17 drops are equal to 1 mL, but you should check this for your particular tubing while the column is flowing after you have attached the Mariotte bottle.

- Set up the Mariotte bottle, and attach it to the column so that the bottom of the tubing in the bottle is above the level of the outlet from the tubing attached to the bottom of the column. Lower the column itself so that the top of its bed is below the level of the outlet at the end of the tubing to the fraction collector.

- Let around 10 mL of 10 mM EDTA, 20 mM sodium phosphate, pH 6.0, flow through the column while you have the outlet tubing flowing into a beaker rather than the fraction collector. Run the column until air is bubbling out of the end of the glass tubing in the Mariotte bottle. If air is trapped in the tubing extending from the reservoir to the column, remove it using the three-way valve and syringe (Figure 1.25).

- At this point, determine the flow rate, and adjust it to around 1.5 ± 0.5 mL min^{-1} by raising or lowering the Mariotte bottle.

- Put 20 tubes into the fraction collector. Close the stopcock at the bottom of the column.

Loading and running the column (40 min).

- Stop the flow of buffer from the reservoir and close the stopcock at the bottom of the column. Move the column up on the rack until the top of the bed is at the height of the end of the tubing in the Mariotte flask so that you maintain the flow rate. Remove the cap from the column and reopen the stopcock at the bottom. Allow buffer to run out of the column until the meniscus is 2–3 mm above the surface of the AMP–agarose beads. Do not let the column run dry! Close the stopcock.

- After you have checked to make sure that the extract does not contain precipitated material, layer the tissue extract on the surface of the beads carefully, to avoid stirring up the solid phase, and open the stopcock.

- As you load the column, you will be collecting the fluid that passes through it in the tubes you have put into the fraction collector. You will collect this fluid by manually advancing the fraction collector. Keep an eye on the tube that is under the outlet and receiving the fluid flowing through the column because if anything goes wrong, the LDH that you are trying to purify may be flowing through the column without associating with it, and you do not want to lose your LDH. When a tube is about two-thirds full, advance the fraction collector to the next tube manually by pushing the advance button. Advance as many times as necessary to capture all of the fluid in successive tubes.

- When most of the extract has entered the solid phase but the meniscus is still 2–3 mm above the top of the solid phase, carefully fill the remainder of the column with buffer.

- Close the bottom stopcock and reattach the cap and tubing extending from the reservoir. When everything is reattached, lower the column so that the level of the top of the solid phase is just below the outlet to the fraction collector, to ensure that the column will never run dry.

- Open the bottom stopcock, and begin collecting fractions of 5 mL automatically.

- Collect six fractions (30 mL). This volume of wash should remove all unbound protein from the solid phase in the column.

- Close the bottom stopcock, lift the column so that the top of the bed is at the level of the bottom of the tube in the Mariotte bottle, and disconnect the reservoir.

Elution of LDH from the AMP–agarose using the adduct of NAD⁺ and pyruvate (1 h).

- Open the stopcock, and allow the remaining buffer to flow out of the column into the next tube in the fraction collector until the meniscus is 2–3 mm above the packed beads. Stop the flow at this point by closing the stopcock, and move to the next tube in the fraction collector.

- Fill the column above the beads with the 0.1 mM solution of the adduct of NAD⁺ and pyruvate. Change the setting on the fraction collector so that subsequent fractions will be 1 mL each.

- Replace the solution in the reservoir with the solution of the adduct, reconnect the cap of the column, and lower the column so that the top of the solid phase is just below the level of the outlet to the fraction collector.

- Open the stopcock. As buffer flows through the agarose, the machine should be collecting 1 mL fractions. As the active sites are occupied by the adduct, the molecules of LDH will no longer be able to associate with the solid phase.

- Continue the elution until 30 mL of the solution containing the adduct has passed through the column. Store the fractions at 4 °C until you are ready to assay them for enzymatic activity.

PART 2: IDENTIFY AND POOL THE PEAK FRACTIONS, AND REMOVE THE ADDUCT

Assay for the presence of LDH in the fractions (1 h).
Assay for activity of LDH as described in LDH Exercise 2.

- Assay the 1 mL fractions in the order in which they were collected. Use a cocktail for these assays that is prepared in a large enough volume to assay 10–15 samples in succession. Prepare the cocktail by adding all of the components, so that the addition of 5 μL of enzyme to each assay will bring the concentration of ingredients to their required final values. Start by testing 5 μL of each fraction. If the change in absorbance for any fraction is greater than 0.2 min^{-1}, dilute appropriately and assay 5 μL of the diluted fraction.

- After you have found the peak of the enzymatic activity and the proper dilution for this most active fraction, use the same dilution for each of the fractions to be assayed. To make the dilutions of the samples from the fractions, take 10 μL from a fraction and mix it with the appropriate volume of buffer in a test tube so that you can take 5 μL of this diluted mixture to perform the assay. After you have performed the assay, pour out the diluted mixture from the tube in which you made the dilution, add an appropriate volume of the buffer, and add 10 μL from the next fraction to be assayed to that test tube. Do not use a new test tube for each dilution; use the same one over and over. Do not worry if the absorbance change for a fraction is less than 0.1 min^{-1}; do not reassay at a lower dilution to get an accurate reading because you will probably be discarding that fraction anyway. Do not perform duplicate assays; you are simply trying to decide what to pool. You will characterize the activity more carefully after the fractions have been pooled and the adduct of NAD$^+$ and pyruvate has been removed from them.

- Also assay the last three of the 5 mL fractions that were collected before the adduct was applied to the column. These samples should contain proteins from the heart that, unlike LDH, have no affinity for the AMP on the agarose, but you want to make sure that this is the case. If there is any LDH in these tubes, assay some more of the 5 mL tubes to make sure that they do not contain a significant amount of LDH.

- Prepare a table of all of the measurements of dA/dt, corrected for the dilutions, and make a graph of these data, plotting dA/dt as a function of fraction number. It is not necessary to convert the values to the units of micromoles minute^{-1} milliliter^{-1} because only the relative activity is needed to identify the peak of enzymatic activity.

Pooling of the affinity-purified LDH (10 min).

- Keeping the pool cold on ice, pool the fractions containing the majority of the activity of LDH eluted with the adduct. Do not pool too many fractions. It is better to lose a small fraction of the enzyme than have too large a volume. You should get used to making uncomfortable decisions like this one. Measure the volume of the pool.

Ultrafiltration under centrifugal force (15 min).

Although the presence of the adduct of NAD$^+$ and pyruvate did not prevent you from assaying *relative* LDH activity in the fractions eluted from the column, the adduct is an inhibitor of enzymatic activity, and the assay is more likely to be accurate if the adduct has been removed and the pH of the pool brought back to 7.5. Removal of the adduct is also essential for obtaining an accurate measure of the concentration of protein by measuring absorbance at 280 nm in the next exercise.

Use ultrafiltration under centrifugal force as described in FNR Exercise 3 (Figure 1.16) and LuxG Exercise 6 to lower the concentration of the adduct ~100-fold from the original concentration of 0.1 mM. Use each ultrafiltration device only once because the pores get clogged.

- Load the pool of purified LDH into one or two Amicon Ultra-4 devices on ice.

- Spin at 7000 rpm in the JA-14 rotor (7500 $\times$ g) at 4 °C until the volume of retentate is about 10% of the original volume. This should take about 2 min.

- Recover the retentate and dilute it 10-fold with 10 mM EDTA, 50 mM sodium phosphate, pH 7.5.

• Transfer the diluted retentate to a fresh Amicon Ultra-4 device (or two devices), then repeat the previous two steps. This should result in an approximately 100-fold dilution of the adduct and the LDH should be at its original concentration.

• Save all of the filtrate that passed through the apparatus in the last ultrafiltration under centrifugal force. It contains all small molecules that absorb at 280 nm, mainly the adduct, at the same concentration that they are in the retentate. Consequently, this diluted filtrate is the ideal blank to use when taking the absorbance of your LDH at 280 nm because it contains everything that is in the diluted retentate except the protein. Dilute this filtrate tenfold with 10 mM EDTA, 50 mM sodium phosphate, pH 7.5, just as you did the retentate. Label it "filtrate for blank" and store it in the cold room.

PART 3: ASSAY THE ENZYMATIC ACTIVITY OF THE PURIFIED LDH AND PREPARE ALIQUOTS FOR STORAGE (40 MIN)

• Assay the diluted retentate from the ultrafiltration for enzymatic activity as described in LDH Exercise 2 before you freeze the rest of it. You should be able to predict what dilution of this solution should give the ideal rate for the enzymatic reaction, but you may have to perform assays until you are in the proper range by making dilutions as necessary. Repeat the assay until you get the same rate in successive triplicates. This determination of enzymatic activity should be as accurate as you can make it.

• Convert the dA/dt to the concentration of enzymatic activity in units of micromoles minute^{-1} milliliter^{-1} and then calculate the total activity of the sample in units of micromoles minute^{-1}.

• Divide all of the remainder of the pool of purified enzyme into 16 equal aliquots.

• Add glycerol to 10% (three drops of glycerol for every milliliter of solution) to each aliquot. Label these portions as "purified LDH." From now on, this solution will be given the name "purified LDH." Freeze all of these 16 aliquots in a slurry of dry ice in isopropanol.

Uses intended for the aliquots of purified LDH:

There are seven types of experiments you may be scheduled to perform with the stored aliquots of purified LDH. All of the LDH exercises that follow this one will require aliquots of purified LDH. The isoenzymatic electrophoresis that was an earlier exercise in the manual may also follow this one on your particular schedule and aliquots of purified LDH from heart may be included in this exercise. The sample was split into 16 aliquots before freezing because some of these experiments require more enzyme than others. The chromatography by molecular exclusion requires the most. This large number of aliquots will make it less likely that any of them will need to be refrozen and then thawed again for another exercise. Multiple cycles of freezing and thawing can cause loss of enzymatic activity.

Analysis of the Data and Questions to Address in Your Report

1. While you were working, you made a graph of the relative enzymatic activity in the fractions from the affinity adsorption and elution to identify the peak of activity for LDH. You have also quantified more accurately the total activity in the pooled fractions from the column after removing the adduct between NAD$^+$ and pyruvate.

2. Consider the materials used for the fractionation with ammonium sulfate and affinity adsorption and elution. Which technique requires the more specialized reagents? Which of these two techniques is more easily adapted to purification of proteins other than LDH?

3. Provide at least two reasons for performing fractionation with ammonium sulfate prior to the affinity selection.

4. What aspect of the structure of LDH is exploited in this purification procedure? Discuss whether or not one of the dyes, Cibacron Blue or Procion Red, suggested for affinity adsorption of FNR earlier in this book might work for adsorption and affinity elution of LDH.

5. You used both AMP and the adduct of NAD^+ and pyruvate as ligands to purify mainly the B_4 isoenzyme of bovine LDH from bovine heart (Kaplan et al., 1974). Discuss the difference in the interactions of the enzyme with the adduct and AMP. What is the advantage of these differences for the affinity selection?

6. Unmodified NAD^+ and NADH can associate with the active site of LDH. Why is it far better to use an adduct of NAD^+ and pyruvate rather than unmodified NAD^+ or NADH to elute this enzyme from the AMP–agarose?

LDH EXERCISE 6

Concentration of Protein by the Bradford Stain and Absorbance at 280 nm

Technical Background and Experimental Plan

The concentrations of the enzymatic activity of LDH in the samples from the various stages of the purification have been assayed. The concentration of protein in each of these samples will now be determined using the Bradford assay that was described in detail in FNR Exercise 2. The concentration of protein in the LDH purified by adsorption and elution will be determined by measuring its absorbance at 280 nm as described in detail in LuxG Exercise 6. The advantage of the former method is that it estimates the concentration of the total protein in a mixture of proteins; its disadvantage is that, even when it is carefully performed, it is not so accurate as the latter. The advantage of the latter method is that it measures a defined property of the protein, namely its known composition of tryptophan and tyrosine, but the disadvantages are that the protein must be pure and there can be no other material in the sample that can interfere with the measurement by absorbing at 280 nm. For example, there can be none of the adduct of NAD^+ and pyruvate in the solution because it absorbs strongly at 280 nm.

The adduct of NAD^+ and pyruvate does not interfere with the Bradford assay, so the assay could have been performed on an aliquot taken before the ultrafiltration under centrifugal force. There is, however, some loss of LDH upon ultrafiltration so you should wait to perform the Bradford assay until you have the pool that was submitted to the ultrafiltration. The measurement of A_{280} must be performed on an aliquot taken after the ultrafiltration. To compare the results of the Bradford assay and the measurement of A_{280}, you should perform the two assays on the same aliquot of the purified LDH, and it must be an aliquot removed after the ultrafiltration under centrifugation.

After making these measurements of the concentrations of protein, the specific activity of LDH can be expressed in the units of micromoles minute^{-1} milligram^{-1} (Boyer, 2012; Ninfa et al., 2010). A table summarizing the purification will be prepared to record the degree of purification of LDH and the percent recovery at each step. The catalytic constant for the purified LDH will then be calculated.

Reagents, Supplies, and Equipment

Aliquots of LDH at various stages of purification:

Initial supernate, 39% ammonium sulfate supernate, 39% ammonium sulfate pellet, 68% ammonium sulfate supernate, 68% ammonium sulfate pellet, dialysate of the protein before the adsorption and affinity elution, and purified LDH.

The "filtrate for blank" from the second ultrafiltration step after affinity purification.

Standard protein for the Bradford assay:
Bovine immunoglobulin G, 0.1 or 1 mg mL^{-1}.

Reagent for Bradford stain:
0.01% (w/v) Coomassie Brilliant Blue G-250, 4.75% (v/v) ethanol, 8.5% (v/v) phosphoric acid.

Buffered solution for dilutions and blanks:
10 mM EDTA, 50 mM sodium phosphate, pH 7.5.

Supplies and equipment for spectrophotometry:

> For the Bradford assay, as in FNR Exercise 2, there are three choices of spectrophotometers and their respective vessels for the samples.

> For the A_{280} assay, as in LuxG Exercise 6, there are two choices of spectrophotometers and sample vessels.

Precaution

The Bradford reagent contains phosphoric acid that can erode flesh and Coomassie Brilliant Blue G-250 that is toxic. Wear gloves.

Before You Come to the Laboratory

Look up the composition of amino acids of the B isoenzymatic subunit of LDH from *Bos taurus*. Using Equation 2.3 in LuxG Exercise 6, calculate its molar extinction coefficient at 280 nm, ε_{280}. Use the UniProt database http://www.uniprot.org for the composition of amino acids. The B isoenzymatic subunit of LDH is the isoenzymatic subunit that predominates in bovine heart. You will be calculating the molar concentration of active sites of the enzyme from the measurements, and there is one active site on each subunit. The fact that the enzyme is a tetramer is irrelevant to this calculation. Make sure that you find the composition of amino acids for only the monomer.

Procedure (One Laboratory Session)

Part 1: Bradford assay of each fraction from bovine heart (1 h).

- Remove from the freezer one aliquot of each of the fractions from the purification of LDH: initial supernate, 39% ammonium sulfate supernate, 39% ammonium sulfate pellet, 68% ammonium sulfate supernate, 68% ammonium sulfate pellet, dialysate of the protein before the adsorption and affinity elution, and purified LDH. Get the filtrate for the blank from the cold room.

- As before, the volumes of the samples placed in the various devices are 0.2 mL for a multi-well plate, 0.5 mL for a cuvette in a conventional spectrophotometer, and 3 mL for a tube in a Spec 20.

- Prepare dilutions of bovine immunoglobulin G and determine a standard curve, as described in the FNR Exercise 2.

- Assay each of the fractions from the purification of LDH as described in FNR Exercise 2. Start with the following dilutions:
  > 1:500 for all fractions except the final purified enzyme.
  > 1:50 for purified LDH.

- Perform the Bradford assays. For each assay mix one part of each diluted fraction plus nine parts staining solution.

- If the A_{595} for a particular sample is not within the range of the absorbances for the standards, adjust the dilution and try again.

Part 2: Absorbance at 280 nm of purified LDH (30 min).

- Perform this measurement as described in LuxG Exercise 6, except that you will assay the purified LDH. Use a sample volume of 0.2 mL for a multiwell plate or 0.5 mL for a cuvette in a conventional spectrophotometer. Use as a blank for this measurement the "filtrate for the blank." Make sure that you have diluted the filtrate by exactly the same factor that you used to dilute the retentate. Subtract the A_{280} of the blank from the A_{280} of the solution of purified LDH.

- If the A_{280} corrected with the blank is greater than 1, make a dilution of the purified LDH until you get a value between 0.3 and 0.8. Dilute the blank just as you dilute the purified LDH. After you have determined the absorbance, the sample can be refrozen and added later to the sample for the chromatography by molecular exclusion.

Analysis of the Data and Questions to Address in Your Report

1. Using the standard curve of immunoglobulin G, calculate, as described in FNR Exercise 2, the concentration of protein in each of the undiluted fractions that you assayed with the Bradford assay. In other words, correct your measurements for the dilutions you performed. Calculate also the total mass of protein in each fraction.

2. Look up the molar mass for a subunit of LDH at http://www.uniprot.org. Using the molar mass for a subunit of LDH and the concentration of protein that you determined by the Bradford assay, calculate the molar concentration of active sites of LDH in units of micromolar in the undiluted solution of purified LDH.

3. Using A_{280} data and the extinction coefficient for LDH at 280 nm that you calculated before you came to the laboratory, calculate the molar concentration of the purified LDH.

4. Compare the two values for the molar concentration of active sites in the solution of purified LDH. Which value is the more accurate for the concentration of the enzyme? If there is a significant discrepancy, suggest an explanation.

5. Specific activity is discussed in FNR Exercise 2. Use the concentration of enzymatic activity for LDH and the concentrations of protein from the Bradford assay to calculate the specific activity of LDH for each fraction in units of micromoles minute^{-1} milligram^{-1}.

6. Prepare a summary of the purification of LDH from bovine heart using the template of Table 3.3. Fold purification is the increase in specific activity. The percent recovery is the yield of total enzymatic activity (μmol min^{-1}) in that fraction relative to the total enzymatic activity in the initial supernate. Before you calculate percent recovery, correct each of the later volumes of solution for the amounts that you removed from the earlier volumes of solution. This is not a simple calculation. First, you have to calculate what the total activity of each fraction would have been before you removed the samples. Second, you have to increase each of the values at each stage after the initial supernate by a factor that takes into account the material you removed from the earlier solutions that consequently did not make it through to the next step. For example, if you removed 15% of the initial supernate for later use, you have to divide the total activities in each of the ammonium sulfate fractions by 0.85 to account for the material you removed from the supernate that otherwise would have been submitted to

TABLE 3.3   Summary of the Purification of Bovine Heart LDH

	Total Volume (mL)	Total Protein (mg)	Concentration of Protein (mg mL^{-1})	Total Activity (μmol min^{-1})	Concentration of Activity (μmol min^{-1} mL^{-1})	Specific Activity (μmol min^{-1} mg^{-1})	Fold Purification	Recovery (%)
Initial supernate								
39% pellet								
39% supernate								
68% pellet								
68% supernate								
Preadsorption dialysate								
Purified by affinity elution								

precipitation. If you removed 11% from the redissolved ammonium sulfate pellet before you dialyzed it, you have to divide the total activity in the dialysate by a further factor of 0.89. If you removed 7% of the dialysate before you applied it to the adsorbent, you have to increase the total activity in purified LDH by an additional division by 0.93. These corrections are to be made only in calculating percent recovery, which is a number that assumes that you carried out the purification without removing any samples along the way.

7. Use the data for the purified LDH to calculate the catalytic constant for the enzyme in the standard units of second^{-1}. The catalytic constant is the rate constant k_0 in Equation i.1. When concentrations of reactants are high enough to saturate the active site of the enzyme, the initial velocity of the enzymatic reaction is equal to $k_0 [E]_t$. You can convince yourself of this fact by taking the limit of Equation i.1 at infinite concentration of reactant A. For the enzymatic assays you performed, the concentrations of NADH and pyruvate were at saturating levels, so the catalytic constant k_0 is the initial velocity of the enzymatic reaction for the purified enzyme in units of micromoles minute^{-1} milliliter^{-1} divided by the molar concentration of active sites. You have calculated the concentration of active sites in units of micromolar, that is micromoles liter^{-1}. As always, make sure the units cancel and you are left with a number in the proper units. In fact, you don't really need to know what to divide by what as long as you start with units of micromoles minute^{-1} milliliter^{-1} and finish with units of seconds^{-1}. The rate constant k_0 is equal to the rate constant for the enzymatic reaction starting from the fully occupied active site and ending with the dissociation of the last product, because at saturation, the reactants associate with the active site so rapidly that, by definition, their association has no effect on the rate of the enzymatic reaction.

8. Compare your calculation of the catalytic constant to those in the literature. In addition to published values for the enzyme from bovine heart, it is worthwhile to compare the catalytic constants for LDH from other mammalian species and tissues. Be aware that in some of the older articles, the catalytic constant is referred to as k_{cat}.

9. The specific activity of LDH at each step of the purification is an indication of the degree of purity. Did the fractionation with sulfate ion and the adsorption and affinity elution both increase the specific activity? Comment on which of these two techniques had the greater impact and explain the reason for the difference.

LDH EXERCISE 7

Electrophoresis of Native Proteins from Bovine Heart and Purified LDH on a Gel of Polyacrylamide

Technical Background

Electrophoresis of a native protein on a gel of polyacrylamide is a stringent test of the purity of that protein (Kyte, 2007, pp 32–47; Ninfa et al., 2010). The migration of a native protein through a matrix of polyacrylamide is a function of both its charge at the pH of the surrounding solution and the resistance to its migration, or sieving, produced by the polymers of polyacrylamide in the separating gel (Figure 3.17). Each protein responds to the pH and the percentage of acrylamide in the gel in its own characteristic way. The electrostatic force produced by the electric field is directly proportional to the charge on the protein, which is a function of the pH of the solution (Figure 1.5). The hindrance to the movement of the protein through the strands of polyacrylamide is an exponential function of its surface area; the greater the surface area, the stronger the hindrance to its movement (Kyte, 2007, pp 423–431).

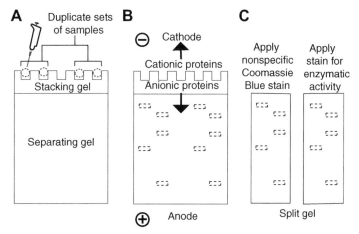

FIGURE 3.17   Separation of native proteins by electrophoresis on a gel of polyacrylamide and localization of a particular enzyme by staining for its activity.   (A) Samples of native proteins are added to the wells in the stacking gel. (B) When the voltage is applied, proteins that are cationic at the pH of the buffer in the upper reservoir migrate upwards and are lost; proteins that are anionic at the pH of the buffer in the upper reservoir migrate through the stacking gel and into the separating gel, in which they are separated one from the other. (C) The separating gel is split in two, and one half is stained for all of the proteins with Coomassie Brilliant Blue and the other half is stained for enzymatic activity.

The protein remains in its native conformation during the electrophoresis, so it retains its biological activity. Therefore, a band of protein identified by a nonspecific stain such as Coomassie Brilliant Blue on one lane of a gel can be assayed for a specific enzymatic activity on another lane that has not been stained with Coomassie Brilliant Blue, which denatures the protein. Thus, it can be determined whether or not the most abundant protein in the preparation represents the enzyme of interest.

Because each protein responds to the pH of the solution and the sieving of the polyacrylamide in its own characteristic fashion, running the sample at two different values of pH or two different concentrations of polyacrylamide provides almost unequivocal proof that the enzymatic activity is associated with the major band of protein and the major band of protein represents only one particular molecule of protein.

In the electrophoresis that you will perform, the anode is at the bottom of the gel and the cathode at the top so only anionic proteins enter the gel. Consequently, you will never see any of the proteins that are cationic at the pH of the solution in the upper reservoir because they will move upward through this solution. For this reason, electrophoresis of this type cannot perform a complete census of all of the proteins in a particular preparation. Electrophoresis of proteins unfolded by dodecyl sulfate, however, because all of the polypeptides are anionic, does perform such a census. Because electrophoresis of the native protein identifies the particular protein that has the enzymatic activity and because electrophoresis of polypeptides saturated with dodecyl sulfate performs a complete census of all of the polypeptides in the sample, these two methods of electrophoresis are complementary to each other.

The charge on a native protein during the electrophoresis is determined by its titration curve (Figure 1.5) and the pH of the solution through which it passes as it moves through the polyacrylamide. The pH in the separating gel, however, is not the same as the pH of the solution used to pour the gel because the stable moving boundary that passes through the gel ahead of the proteins lays down a solution of a composition that is different from the original solution. Proteins with a pI below the pH in the separating gel are anionic, those with a pI above the pH in the separating gel are cationic. The stacking system chosen for this experiment lays down a solution with a pH greater than the pI of LDH so LDH moves continuously through the separating gel. Both the tertiary structure and the quaternary structure of each protein are maintained because no denaturants are used.

If you have performed LDH Exercise 3, you have already performed electrophoresis of native LDH. If you have, you have already learned that electrophoresis of native LDH works well (Dietz and Lubrano, 1967; Opher et al., 1966). The stain that will be used for the activity of LDH in this exercise is the same one that was used in LDH Exercise 3 (Figure 3.6). The composition of the gel of polyacrylamide will be similar to that used in FNR Exercise 4, FNR Exercise 7, and LuxG Exercise 3. Stacking with discontinuous stable boundaries will again be used (Davis, 1964; Ornstein, 1964; Tulchin et al., 1976). The proteins, however, will not be unfolded with dodecyl sulfate, which would destroy their native structure. The concentration of acrylamide in the separating gel will be 8% in this case, which is different from the concentrations in the other exercises.

Experimental Plan

This exercise may be performed over two laboratory sessions but it can be performed in one, especially if you have already run electrophoresis on gels of polyacrylamide and are adept at the procedure. If you are scheduled to perform it in two sessions, the mold is assembled and a solution of acrylamide for the lower separating gel is poured into the mold and polymerized in the first session. Samples of the initial supernate from the homogenate of the heart and the purified enzyme are also prepared for the electrophoresis in this session. In the second session, an upper stacking gel of a lower concentration of acrylamide is prepared, poured on top of the separating gel, and polymerized. Samples are then loaded for electrophoresis in the cold room. As in the exercise in which the isoenzymes of LDH were separated, it is critical to keep the temperature low during the electrophoresis to avoid inactivation of the enzyme by the ohmic heating of the current.

Reagents, Supplies, and Equipment

Samples to be assayed:

Initial supernate from the homogenate of the heart and purified LDH.

80% glycerol, 0.25% Bromophenol Blue.

Stock solutions for the stain for enzymatic activity:

3 M lactate.

75 mM NAD$^+$.

10 mM 5-methylphenazinium methyl sulfate.

3.7 mM Nitro Blue Tetrazolium chloride.

Simply Blue (Invitrogen), which is Coomassie Brilliant Blue G-250 in an aqueous solvent.

20% NaCl: to be added to the Simply Blue for overnight staining.

Whatman filter paper.

Plastic trays.

All of the reagents for preparation of the gels of polyacrylamide and the electrophoresis, and the apparatus and power supply for the electrophoresis listed in FNR Exercise 4. The only difference is that dodecyl sulfate is not included in any of the solutions for the electrophoresis, in any the solutions used to make the samples, or in any other solution used in this exercise.

Precautions

Acrylamide is a neurotoxin that can be adsorbed through the skin: wear gloves and use pipette bulbs. Never pour acrylamide down the drain. Polymerized acrylamide is much less toxic. Polymerize any unused acrylamide and then toss the solid into the trash.

Be careful around the electrophoretic apparatus, which poses the hazard of electric shock; accidentally sticking a finger into the reservoir of a running gel can produce a fatal shock.

Although the Simply Blue stain does not contain organic solvents, the Coomassie Brilliant Blue in the solution is toxic so wear gloves when handling it and leave the trays for the overnight staining in the fume hood.

Before You Come to the Laboratory

You should have already calculated the concentration of enzymatic activity in units of micromoles minute^{-1} milliliter^{-1} and the concentration of protein in units of milligrams milliliter^{-1} for the initial supernate and the purified LDH. Before you come to the laboratory, refer to the subsection Preparing the separating gel and the samples for electrophoresis (below). Calculate any dilutions that will be necessary and the volume of each sample that will contain the appropriate mass of protein (mg) or total enzymatic activity (μmol min^{-1}) for each lane. Write down precisely how you will prepare the dilutions before you come to the laboratory. You must come to the laboratory prepared with all of your data and calculations done in advance, or you will not finish on time.

Procedures

PART 1: PREPARING THE SEPARATING GEL AND THE SAMPLES FOR ELECTROPHORESIS

Preparing the mold for the gel (10 min).

- Assemble the mold for the gel as described in FNR Exercise 4. Be certain the glass plates are clean and dry before assembling.

Preparing the mixture of 8% acrylamide for the separating gel (10 min).

- Mix in a flask, by swirling, the following ingredients in the following order:

30% acrylamide, 0.8% bisacrylamide	8 mL
1.5 M Tris chloride, pH 8.8	7.5 mL
Water	14.2 mL
10% ammonium persulfate	300 μL
Tetramethylethylenediamine	15 μL
Total	**30 mL**

Before adding the tetramethylethylenediamine and the ammonium persulfate to the mixture containing the acrylamide, make sure that you have everything assembled and ready for pouring the gel. The solution of acrylamide must be poured between the plates immediately after adding the catalysts.

Pouring and polymerizing the separating gel (40 min).

- After pouring the mixture containing the acrylamide into the mold, overlay the solution in the mold with 95% ethanol.

- After the gel has polymerized, pour the ethanol off of the top and rinse the space above the gel several times with water.

- If the gel is to be stored overnight, cover it with 0.375 M Tris chloride, pH 8.8 (1.5 M Tris chloride, pH 8.8, diluted fourfold with water), and seal the top of the mold with tape. Store at 4 °C.

Preparing samples of enzyme for electrophoresis while the gel polymerizes (1 h).

- While the gel is polymerizing, remove from the freezer one aliquot each of the initial supernate from the homogenate of bovine heart and purified LDH. Thaw them and put them on ice.

- Each of the two solutions of enzyme that you just thawed must be assayed for enzymatic activity before preparing the samples for the electrophoresis, to check that they are still as active as they were when you froze them. You do not have to perform the assay in triplicate, just until you are sure the LDH in the two solutions is still alive. If either has lost significant activity, correct your calculations for the appropriate lane for the loss in enzymatic activity. Rely on the Bradford assays you have already done for the concentrations of protein because they cannot be affected by freezing.

- The following amounts of each sample are to be loaded on each of the 10 lanes of the gel:

Lanes 1 and 6:	25 μg of purified LDH.
Lanes 2 and 7:	10 μg of purified LDH.
Lanes 3 and 8:	0.2 μmol min^{-1} of purified LDH.
Lanes 4 and 9:	50 μg of supernate from the homogenate of heart.
Lanes 5 and 10:	0.05 μmol min^{-1} of supernate from the homogenate of heart.

Prepare enough of each sample for four lanes, although the plan is to load only two lanes for each. The remainder is backup in case there is a problem with loading. Prepare the samples for each pair of lanes, by mixing:

1. Enzyme: four times the amount desired for each lane in as small a volume as possible but a volume greater than 20 μL.
2. 80% glycerol, 0.25% Bromophenol Blue: one-half the volume of enzyme.
3. 0.5 M Tris chloride, pH 6.8: one-half the volume of enzyme.

It is convenient if the final volume is easily divided by four.

- When you are ready to do so, one-fourth of each sample will be loaded into one lane and another fourth of each into the other of the respective pair of lanes.

PART 2: PREPARE THE STACKING GEL, PERFORM THE ELECTROPHORESIS, STAIN FOR ENZYMATIC ACTIVITY AND PROTEIN

The preparation of the stacking gel (4.5% acrylamide) and the electrophoresis are performed as described in FNR Exercise 4, with the following exceptions:

1. The buffered solutions used for the stacking gel and in the reservoirs do not contain dodecyl sulfate.

2. Electrophoresis is performed at 4 °C to prevent denaturation of enzyme by the ohmic heating generated by the current.

3. The time of the run is longer.

4. After the electrophoresis, the separating gel is split down the middle so that the two different techniques for staining may be applied.

Preparing the stacking gel (30 min).

The preparation of the stacking gel must be done on the same day that the electrophoresis is to be run. Otherwise, the two buffer systems will equilibrate with each other across the interface between the stacking gel and the separating gel, and the discontinuous buffer system required to jump the pH and release the proteins from the moving boundary will fade away overnight.

Electrophoresis (2.5 h).

- Load the samples, overlay them, and fill the reservoirs with 0.192 M glycine, 0.025 M Tris, as described in the instructions for FNR Exercise 4.

- Start with the power supply at 125 V until the Bromophenol Blue enters the separating gel (~40 min), then increase the voltage to 250 V.

- Continue electrophoresis until the Bromophenol Blue is within 1 cm of the bottom of the separating gel (~1.5 h at the higher voltage). Total run should be around 450 V h.

- Turn off the power supply.

- Take the apparatus to the sink, and pour the buffer out of the reservoirs.

- Remove the sandwich of gel and glass plates from the apparatus, and lay it on your bench.

- Using a plastic spacer or spatula, slowly and steadily pry the glass plates apart from one corner. Be ready to catch the plate quickly if it pops up so as not to have it slide and shear the gel. Discard the stacking gel; all of the proteins that did not move toward the cathode will be in the separating gel.

- While the separating gel is still on the bottom plate, cut it in half between lanes 5 and 6 with a fresh razor, not a spatula. The cut must be clean.

Staining for total protein (40 min or 20 min followed by an overnight step).

- Wearing gloves, carefully pick up the half of the gel containing lanes 1–5 from its bottom corners and place it into a tray of water and stain these lanes for total protein with Simply Blue. Follow the instructions in FNR Exercise 4. This gel is 8% acrylamide, so either the microwave or overnight method will work. If using the microwave method, wait to put it in the microwave until the staining for enzymatic activity is running. The microwave method will require 40 min in this laboratory session whereas the overnight method will require 20 min in this session followed by an overnight step.

Staining for enzymatic activity at room temperature (30 min).

- While one partner is opening the mold and cutting the gel, the other partner should mix the ingredients for the stain for the activity of LDH (Table 3.4). In addition to mixing the stain, this partner should also obtain a piece of Whatman filter paper of approximately the same size as the half of the gel containing lanes 6–10.

TABLE 3.4   Mixing the Four Stock Solutions for the Stain

Component	Volume (mL)	Final Concentration (mM)
3 M lactate	1.4	378
75 mM NAD+	2	14
10 mM 5-methylphenazinium	0.7	0.63
3.7 mM Nitro Blue Tetrazolium	7	2.3

- Pour the mixture of dyes and substrates into a plastic pan and submerge the filter paper so that it absorbs the mixture.

- Lay the soaked strip of filter paper over lanes 6–10 of the gel. Be certain there are no air spaces between the filter and the gel.

- Leave the filter paper in contact with the gel at room temperature until distinct blue bands develop. Strong signals will be visible through the Whatman paper. To see weaker signals, it will be necessary to look from below. Bands of stain should appear in 5–10 min.

- When the stain has developed so that the bands are sharp and dark, but before the stain spreads out away from the centers of the bands, remove the Whatman paper.

- While the gels are staining, clean all the pieces of the mold with water and detergent and rinse the reservoirs with water.

Capturing images of the gels (15 min).

- Make images of the gels as soon as possible. Refer to instructions in FNR Exercise 4. Remember to have a plastic ruler along the edge of each gel.

Analysis of the Data and Questions to Address in Your Report

1. Examine the image of the lanes stained with Coomassie Brilliant Blue. Did the affinity adsorption and elution select a single protein? How effective was it at purifying this protein?

2. Examine the image of the lanes that were stained for enzymatic activity. How does the result for the lane containing the supernate of the homogenate demonstrate that the stain is specific for LDH?

3. Compare the image of the lanes stained with Coomassie Brilliant Blue to that of the lanes stained for enzymatic activity. What convinces you that LDH was selected by the affinity adsorption and elution as it was designed to do? Were any other proteins selected?

4. In this electrophoresis of the native proteins, the + receptacle on the power supply is connected to the electrode in the bottom buffer tank, a fact that makes this electrode the anode. How does the result of LDH Exercise 3 assure us that the isoenzymes of LDH in these samples will migrate towards the anode?

5. Why are more volt-hours of electrophoresis necessary for this electrophoretic separation of native proteins that the volt-hours used for separation of polypeptides saturated with dodecyl sulfate in FNR Exercise 4 and LuxG Exercise 3?

6. Discuss whether or not it is likely that all of the proteins in the supernate of the homogenate of the heart entered the separating gel in this electrophoresis.

LDH EXERCISE 8

Electrophoresis of Unfolded Polypeptides from Bovine Heart and Purified LDH Saturated with Dodecyl Sulfate

Technical Background

Refer to the description of the electrophoresis of polypeptides unfolded by dodecyl sulfate on a gel of polyacrylamide in FNR Exercise 4. A critical difference between this technique and the electrophoresis of native proteins on a gel of polyacrylamide is that all of the polypeptides, which are all coated with dodecyl sulfate, enter the gel. In the latter procedure, only the proteins that, at the pH of the buffer in the upper reservoir, have a net charge opposite to that of the electrode at the bottom of the gel migrate into the polyacrylamide.

Experimental Plan

You will compare, by electrophoresis, proteins from the following fractions: (1) supernate of the homogenate of bovine heart, (2) the redissolved pellet from the fractionation by ammonium sulfate that you dialyzed, and (3) the purified LDH. The proteins in these three fractions are unfolded and saturated with dodecyl sulfate by adding sodium dodecyl sulfate and boiling. These samples and a set of standard polypeptides are applied to the gel and separated by electrophoresis. All of the polypeptides in each sample should enter the separating gel and be stained. Thus, an assessment of the effectiveness of the methods for purification may be made by comparing the number and pattern of bands in the various lanes.

Reagents, Supplies, and Equipment

Samples from the freezer of the supernate of the homogenate of bovine heart, the redissolved pellet from the fractionation by ammonium sulfate that you dialyzed, and the purified LDH.

Reagents for unfolding the polypeptides with dodecyl sulfate, for the electrophoresis, for loading the gels, and for staining the gels, as well as all of the supplies and equipment for the electrophoresis, are the same as those in FNR Exercise 4 and LuxG Exercise 3.

Precautions

Acrylamide is a neurotoxin that can be adsorbed through the skin or your digestive tract: wear gloves and use pipette bulbs. Never pour acrylamide down the drain. Polymerized acrylamide is much less toxic. Polymerize any unused acrylamide and then toss the solid into the trash.

Even though it has a safety interlock, you should be careful around the electrophoretic apparatus, which poses the hazard of electric shock; accidentally sticking a finger into the reservoir of a running gel can produce a fatal shock.

Although the Simply Blue stain does not contain organic solvents, the Coomassie Brilliant Blue it contains is toxic so wear gloves when handling it and leave the trays for the overnight staining in the fume hood.

Refer to the figures in FNR Exercise 4 for guidance in preparing and loading the gel and for the process of electrophoresis. Remember to assemble the plates and spacers before you mix the ingredients for the gel of polyacrylamide because the solution must be poured into the mold immediately after adding the catalysts.

Before You Come to the Laboratory

You should have already determined the concentration of protein (mg mL^{-1}) for the supernate of the homogenate of bovine heart, the redissolved pellet from the fractionation by ammonium sulfate that you dialyzed, and the purified LDH. Refer to Table 3.5. Before you come to the laboratory, calculate the amount of each fraction needed to obtain the mass of protein indicated in the table. Dilutions may be necessary. Come prepared to the laboratory with all of your calculations done in advance.

TABLE 3.5   Samples of Fractions from the Purification of LDH and Standards to be Loaded[a]

	Concentration of Protein (mg mL^{-1})	Volume (μL)	Mass (μg)
Standard polypeptides[b]	2	8	16
Supernate of homogenate from heart			10
			30
Dialyzed ammonium sulfate fraction			10
			30
Purified LDH			4
			12

[a] The table may be completed with concentrations of protein from Bradford assays.
[b] The standard polypeptides from Sigma-Aldrich have already been unfolded and saturated with dodecyl sulfate. This sample is ready to load and does not need to be heated.

Procedures (One Laboratory Session)

Preparing the separating gel (40 min).

- Prepare the mold and pour the separating gel of 12.5% polyacrylamide as described in FNR Exercise 4. Everything is the same except that none of the solutions contain sodium dodecyl sulfate.

Preparing the samples of LDH for electrophoresis (30 min).

- While the gel is polymerizing, take the aliquots of the three fractions from the freezer, thaw them, and put them on ice.

- Measure out the amounts of each sample needed for the electrophoresis. You have already calculated the mass of each protein mixture or purified protein to be loaded (Table 3.5). As in the previous exercise involving electrophoresis of native proteins, you should prepare twice this amount of each sample. If all goes well, you will only use half of each sample. It is good, however, to have spare sample if there is a problem loading the gel.

- To each of the three samples, add, in the proper order (Table 1.3), the appropriate amounts of water, buffer, and 2-sulfanylethanol, except you should double the volumes of each to make a final volume of 100 μL. Only 50 μL of each sample, however, will be loaded onto each lane. Calculate to be sure that at least 2 μg of sodium dodecyl sulfate will be added for every microgram of protein in addition to the amount of sodium dodecyl sulfate needed to make its final concentration in the sample greater than 0.1%, and then add the dodecyl sulfate and swirl. Immediately heat the sample in the boiling water for 1 min, then add the Bromophenol Blue and the glycerol. After they are prepared, the samples can sit for as long as necessary.

Preparing the stacking gel, loading the samples, and electrophoresis (2.5 h).

- Follow the instructions in FNR Exercise 4.

Staining, and making the image of the stained gels.

- Follow the instructions in FNR Exercise 4. For staining, either the microwave or overnight method may be used.

- Timing:
  For the microwave staining method, 40 min will be necessary; for the overnight staining method 20 min will be necessary during this session followed by an overnight soak.

Analysis of the Data and Questions to Address in Your Report

1. Prepare the plot of the negatives of the logarithms of the relative mobilities of the standard polypeptides as a function of their length, as you did in FNR Exercise 4.

2. Estimate the length of the polypeptide constituting LDH by interpolation on the plot of the negatives of the logarithms of the relative mobilities (R_f). Is your estimate close to the actual length of the polypeptide?

3. The three fractions examined in this experiment allow you to evaluate the outcome of the fractionation with sulfate ion and the adsorption and affinity elution. Comment on the relative effectiveness of the two techniques.

4. Why was a separating gel of 12.5% polyacrylamide used to examine LDH rather than a separating gel of 15% polyacrylamide, as that used for LuxG?

5. Discuss whether or not it would be possible to assay LDH activity rather than staining with Coomassie Brilliant Blue in this type of electrophoresis experiment.

6. Discuss whether or not the result of this experiment by itself is sufficient to conclude that LDH has been purified to homogeneity by the affinity adsorption and elution.

7. Discuss whether or not this experiment by itself says anything regarding the quaternary structure of bovine LDH as did the electrophoresis of the hybrids of the isoenzymatic subunits of LDH on cellulose acetate.

LDH EXERCISE 9

Determination of the Michaelis Constant, K_{mNADH}, of the Isoenzymes of LDH for NADH

Technical Background

Refer to FNR Exercise 5 and LuxG Exercise 5. There are the following differences in this instance:

1. The cosubstrate is pyruvate rather than iodonitrotetrazolium (INT) or flavin mononucleotide (FMN). The reported Michaelis constant, $K_{m,pyr}$, for pyruvate of the human A_4 isoenzyme of LDH, which is the homotetramer of A isoenzymatic subunits, is 0.16 mM (Eszes et al., 1996), so all reactions will have 1 mM of pyruvate to ensure that this reactant is at a concentration that saturates the enzyme as the concentration of NADH is systematically raised (Equation i.2).

2. The range of concentrations of NADH will be 12.5 to 400 μM.

3. The rate of the reaction catalyzed by LDH is monitored by the decrease in A_{339} as NADH is oxidized to NAD^+, as described in LDH Exercise 2. This is a simple and effective way of quantifying the rate of the enzymatic reaction catalyzed by LDH because neither pyruvate nor lactate absorbs light of this wavelength. One consequence of using this assay is that, unlike in FNR Exercise 5 and LuxG Exercise 5, the initial absorbance of the reactions will vary as the concentration of NADH is varied. Because reaction rates are quantified only by the change in absorbance, however, the variation in initial A_{339} is irrelevant to the measurement of the rate.

4. To convert dA_{339}/dt to the initial velocity, the change in the extinction coefficient, $\Delta\varepsilon_{339}$, for the conversion of NADH to NAD^+ at 339 nm (6.2 mM^{-1} cm^{-1}) will be used. Calculation of this value for the $\Delta\varepsilon_{339}$ was presented in LDH Exercise 2.

5. The reactions will be run in 100 mM sodium phosphate, pH 7.

Experimental Plan

Refer to FNR Exercise 5 and LuxG Exercise 5. There are the following differences in this instance:

1. Two samples from bovine tissues will be assayed: the crude extract from skeletal muscle prepared in LDH Exercise 1 and purified LDH. If you were assigned heart, kidney, or liver for homogenization in LDH Exercise 1, obtain some extract of bovine skeletal muscle from another student in your class.

2. Assays will be performed as described in LDH Exercise 2.

Reagents, Supplies, and Equipment

Bovine skeletal muscle extract and purified LDH on ice (thaw just before use).

40 mM NADH (minimize exposure to atmospheric oxygen).

100 mM sodium pyruvate.

100 mM sodium phosphate, pH 7.5.

500 mM sodium phosphate, pH 7.5.

Assorted pipettes.

Plastic snap-cap tubes (1.5 mL).

Disposable test tubes.

Parafilm.

Micropipettes: 20, 200, and 1000 μL.

Electric timer.

Vortexing mixer.

Supplies and equipment for spectrophotometry:

> If you are using a plate reader, you will need a multiwell plate that is transparent to ultraviolet light (Figure 1.13 and Appendix I).

> If you are using a conventional spectrophotometer, you will need a cuvette that is transparent to ultraviolet light and that has a narrow, tapered chamber for which only 0.5 mL of sample is required.

> If you are using a Spec 20, its glass culture tubes are partially transparent to light of 339 nm.

> A plate reader, a conventional spectrophotometer, or a Spec 20 to quantify A_{339}.

Before You Come to the Laboratory

The 500 mM sodium phosphate is a stock solution that is diluted to achieve a final concentration of buffer of 100 mM in the assays. A solution of 100 mM sodium phosphate is used for dilution of the enzymes. The volume of the assay is determined by the type of reaction vessel and spectrophotometer you will be using: 0.2 mL for a multiwell plate in an automated plate reader, 0.5 mL for a cuvette in a conventional spectrometer, or 3 mL for a glass culture tube in a Spec 20. Decide what the reaction volume will be, then do the following calculations:

- In previous exercises, the concentration of enzymatic activity of LDH in the extract from bovine skeletal muscle and purified LDH was determined. Use these concentrations of enzymatic activity to calculate the appropriate dilutions and volumes necessary to generate a dA_{339}/dt of around 0.15 min^{-1} in the volume appropriate to the instrument you will be using.

- Plan what stock solution you will prepare before you begin the assays so that all you have to do for each assay is to add a particular volume of that stock solution, a particular volume of the diluted NADH, and a particular volume of enzyme, in that order. Determine what these volumes will be.

- Determine how you will dilute the stock solution of 40 mM NADH and what volumes of each of the dilutions you will use to obtain the required concentrations for estimating the Michaelis constant. You will perform assays with final concentrations of NADH of 0, 12.5, 25, 50, 100, 200, and 400 μM. These concentrations are achieved by serial dilution. Read the detailed instructions covering serial dilution in FNR Exercise 5.

Procedure (One Laboratory Session, 2.5 h)

• Thaw one tube each of frozen supernate from a homogenate of bovine skeletal muscle and the purified LDH. You may hasten the thawing process by rubbing the tube between your hands, but put the tube on ice as soon as it is thawed completely.

• Perform the assay at 400 μM NADH, 1 mM pyruvate, first for each of the two samples. Use this assay to adjust the dilutions of protein to get in the proper range of absorbance change if necessary, and then use that dilution of the extract for all of your other measurements at the other concentrations of NADH. It is important that exactly the same amount of LDH is added to each assay; make a large enough volume of the correct dilution so that only one tube of diluted extract can be used for all of your assays at the different concentrations of NADH. Keep the diluted enzyme on ice.

• Perform a reaction with 400 μM NADH but no pyruvate for both samples of LDH. There should be very little decrease in the A_{339} within the first 3 min. If there is, repeat this control. If there is still a significant decrease in A_{339} in the absence of pyruvate you will have to do controls without pyruvate for each assay at the different concentrations of NADH.

• Perform all of the assays as described in LDH Exercise 2 with 1 mM pyruvate and separate assays with concentrations of NADH of 0, 12.5, 25, 50, 100, 200, and 400 μM. You have already determined how you will be making a stock solution for repetitive use and all of the dilutions you need to make.

• For each reaction, add everything except the enzyme to the reaction vessel, check the A_{339}, add the enzyme, and measure the rate. Record the absorbance at 30 s intervals for 3 min.

Analysis of the Data and Questions to Address in Your Report

1. Determine the concentration of enzymatic activity (μmol min^{-1} mL^{-1}) in the supernate from the homogenate of skeletal muscle from the initial velocities you measured for each concentration of NADH using the change in extinction coefficient for the conversion of NADH to NAD$^+$.

2. Determine the Michaelis constant for NADH, K_{mNADH}, for LDH in skeletal muscle by plotting the concentration of enzymatic activity as a function of the concentration of NADH in micromolar and fitting Equation i.2 to the data. The units on the values of K_{mNADH} should be micromolar.

3. Determine the enzymatic activities of the purified LDH in units of seconds^{-1} from the initial velocities you measured for each concentration of NADH. First calculate the concentration of enzymatic activity (μmol min^{-1} mL^{-1}) for each measurement, and then divide that number by the molar concentration of LDH (μmol mL^{-1}) in each assay. This is the same conversion of units that you did when you calculated the catalytic constant for LDH in LDH Exercise 6. In the current experiment, however, the variation in concentration of the reactant NADH generates a range of values in seconds^{-1} rather than a constant.

4. Determine the Michaelis constant for NADH, K_{mNADH}, for LDH purified from bovine heart by plotting the enzymatic activity in seconds^{-1} as a function of the concentration of NADH in micromolar and fitting Equation i.2 to the data. You will be using different units for activity for this plot than were used in point 2 for the plot of LDH from skeletal muscle. You should convince yourself that regardless of which of the two types of units for activity are used for plotting the data, the K_m will be obtained in the units used for the concentration of reactant, micromolar in this case.

5. Compare your results to those previously reported in the literature for LDH from skeletal muscle and heart. When examining the literature, be aware of species differences.

6. Discuss whether or not the calculated K_m convinces you that the enzyme you have purified is, in fact, LDH.

7. Discuss the concentration of NADH that LDH is likely to encounter in its natural environment, the cytoplasm of a cell from bovine skeletal muscle or a cell from bovine heart.

8. If you have determined the Michaelis constant for NAD(P)H for FNR or LuxG or both of these enzymes, compare this value or these values to those obtained for LDH from heart and skeletal muscle. What does this comparison tell you regarding the concentration of NAD(P)H these enzymes are likely to encounter in their natural environment?

LDH EXERCISE 10

Chromatography by Molecular Exclusion to Evaluate the Quaternary Structure of LDH

It is possible, and even strongly recommended, to run both FNR and LDH at the same time on the same column with the same standards, and assay the one set of fractions for the standards and FNR and LDH to locate the elution volume of each protein, rather than running them separately. Wait until samples of both purified FNR and purified LDH have been prepared and only then pour and run the chromatographic column following the directions in FNR Exercise 6. In this case, the column can be run with both FNR and LDH as well as all of the standards in the same sample, and the fractions can be assayed for the activities of the standards, FNR, and LDH all on the same run.

Technical Background and Experimental Plan

Read the discussion of chromatography by molecular exclusion and the experimental plan in FNR Exercise 6.

Reagents, Supplies, and Equipment

Regardless of whether or not you have already performed FNR Exercise 6, the same reagents, supplies, and equipment listed in that exercise will be used, with the following exceptions:

1. A sample of purified LDH with a total activity of 25 μmol min^{-1} will be loaded onto the column, rather than FNR.

2. The reagents for the assay of the activity of LDH, rather than those for the assay of FNR, are needed.

Precautions

Stir the Sephacryl S-200 gently and as little as possible; the solid phase easily breaks up into smaller fragments that interfere with the flow of the buffer through the chromatographic bed.

Once the column has been poured, never allow it to run dry.

Before You Come to the Laboratory

Use your measured concentration of the enzymatic activity of purified LDH to calculate how many frozen aliquots you will need to thaw to have the 25 μmol min^{-1} that is needed.

Procedures (Three Laboratory Sessions)

If you performed the chromatography by molecular exclusion in FNR Exercise 6, you will still have a calibrated column of Sephacryl S-200 in the cold room to use in the present exercise. If you do, the time that will be required for this exercise will be much less than if you are pouring and calibrating a column of Sephacryl S-200. If you have waited, however, until you have samples of both purified FNR and purified LDH before pouring, running, and calibrating the column, you will save even more time.

- If you have a column already, start it flowing at an appropriate time based on the schedule for this course. If the top of the Sephacryl is not level, level it as you did in FNR Exercise 6 before you start it flowing. You should let a volume equal to the bed volume of the column flow through it before you add the sample unless you completed FNR Exercise 6 in the last week.

The flow through the column will probably have to be started the day before you need to use it, but you should decide on when to start it based on its rate of flow. Let it flow into a beaker or flask.

The bed volume of the column should be almost the same as it was when you used it before, so there is no need to recalibrate it with standards and standards do not need to be added to the sample. You should, however, include the blue dextran and the cytochrome c in the sample because they are easy to find. The former controls for any change in the void volume and the latter for any change in the total volume of the column. Because you will not have to assay for the enzymatic activity of the standards again, you will also save time. Load the LDH, the blue dextran, and the cytochrome c, all in the same sample, run the column, and assay the fractions as described in FNR Exercise 6.

- If you have not performed and will not be performing FNR Exercise 6, follow the instructions in that exercise with the following exceptions:

1. Purified LDH from bovine heart (25 μmol min^{-1}), rather than FNR, is mixed with the standards for calibrating the chromatographic column, and the mixture is loaded onto the column.

2. Rather than assaying fractions for the activity of FNR, they are assayed for the activity of LDH as in LDH Exercise 2. Volumes to be assayed from the fractions from the column should be 100 μL for a 3 mL reaction (Spec 20), 30 μL for a 1 mL reaction (conventional UV/vis spectrophotometer), or 10 μL for a 0.2 mL reaction (multiwell plate).

Analysis of the Data and Questions to Address in Your Report

The first three instructions in the section Analysis of the Data and Questions to Address in Your Report from FNR Exercise 6 will guide you through the steps of plotting the assays of the fractions, obtaining values from the internet for the total number of amino acids in each of the standards, preparing a plot of the negative natural logarithms of the partition coefficients ($\alpha_{acc,i}$) for the standards as a function of their total number of amino acids, and interpolating to estimate the number of amino acids in the native oligomer of LDH from bovine heart or estimate the number of amino acids in LDH and the number of amino acids in FNR at the same time.

1. To estimate the number of subunits in the native quaternary structure of an oligomeric protein, both the results from the electrophoresis of the unfolded polypeptides that constitute the protein and those from the current experiment can be combined. What do these results tell you regarding the quaternary structure of LDH from bovine heart?

2. Does this estimate of the number of subunits agree with the conclusion you drew about the quaternary structure of LDH from examining the isoenzymatic hybrids of the enzyme in LDH Exercise 3?

3. Even if the two results agree, assume that they disagree and make a decision about which result would be the more reliable and explain why it is the most reliable.

4. Electrophoresis of a native protein on a gel of polyacrylamide also separates the native molecules of protein by the same process of sieving. How would you design an electrophoretic experiment to estimate the number of amino acids in native LDH?

LDH EXERCISE 11

Detection of LDH in Extracts of Bovine Tissue by Immunoblotting

Technical Background and Experimental Plan

Read the discussion of immunochemistry and the experimental plan for immunoblotting in FNR Exercise 7.

Reagents, Supplies, and Equipment

Assemble the same reagents, supplies, and equipment listed in the FNR Exercise 7, with the following exceptions:

1. Samples of protein from bovine heart and bovine skeletal muscle will be run rather than proteins from the chloroplasts of spinach.

2. Rather than anti-FNR, the primary antibodies will be the following:

   a. Rabbit-anti-LDH-A (affinity purified): Sigma-Aldrich, item SAB2101329.

   b. Rabbit anti-LDH-B (affinity purified): Sigma-Aldrich, item AV48210.

   The anti-LDH antibodies are supplied by this company as dry powders. The directions accompanying the powders recommend resuspending them at 1 mg mL^{-1} in phosphate-buffered saline with Tween-20. The procedure provided below assumes such stocks are being used. If you are provided with stocks at a different concentration or from a different supplier, adjust the procedure accordingly.

Before You Come to the Laboratory

The primary antibodies that will be used for detecting the isoenzymatic polypeptides of LDH, anti-LDH-A and anti-LDH-B, are specific to the unfolded polypeptides of the A and B isoenzymatic subunits, respectively. For immunostaining the polypeptide that folds to form the B isoenzymatic subunit, which predominates in bovine heart, two of the fractions examined in LDH Exercise 8, the initial supernate and purified LDH, will be assayed. If you or other students in your class prepared a supernate from the homogenate of bovine skeletal muscle earlier in the course in LDH Exercise 1, aliquots of this (unpurified) protein mixture will be immunostained for the polypeptide of the A isoenzymatic subunit.

Immunoblotting, under ideal conditions, is able to detect as little as 0.1 ng of a polypeptide. Although you do not need to test this limit, much less protein will be loaded in each of the samples than you loaded on the gel of polyacrylamide in LDH Exercise 8. The amounts that are to be loaded, which are based on the concentrations of protein from the Bradford assays you performed in LDH Exercise 6 or will perform in the first part of the current exercise, are listed in Table 3.6.

- Before you come to class, complete the lines in this table pertaining to samples derived from bovine heart, noting any dilutions of the samples of the fraction that will have to be performed to get reasonable volumes. You will complete the lines pertaining to protein from bovine skeletal muscle after assaying the concentration of protein in this sample as described below.

TABLE 3.6   Samples of Protein to be Analyzed by Electrophoresis Followed by Immunoblotting[a]

Sample	Concentration of Protein (mg mL⁻¹)	Volume (µL)	Mass (µg)
Commercial prestained markers[b]	Unknown	10	Unknown
Initial supernate from heart			2
			10
Purified LDH			0.02
			0.1
Supernate from skeletal muscle			2
			10

[a] The table may be completed using concentrations of protein from Bradford assays.

[b] The volume shown is appropriate for the ColorBurst markers from Sigma-Aldrich. The company does not provide a concentration for the covalently stained proteins in the mixture. These covalently stained polypeptides are already saturated with dodecyl sulfate and are run as provided.

Procedure (Two Laboratory Sessions)

FIRST SESSION

Concentration of protein in the supernate of the homogenate from skeletal muscle (30 min).

You should determine the concentration of protein in the supernate of the homogenate from skeletal muscle. This will help you to load appropriate amounts onto the gel. The homogenization of skeletal muscle and heart should have been done in a similar manner, so the concentration of protein in the supernate of the homogenate from skeletal muscle should not be wildly different from that in the supernate of the homogenate from heart.

- Estimate the concentration of protein in the supernate of the homogenate from skeletal muscle based on the concentration of the supernate from heart, then perform a Bradford assay on the supernate from skeletal muscle.

Preparing the gel of polyacrylamide and the samples of protein, loading the gel, and performing the electrophoresis (3.5 h).

- Pour the separating gel as described in FNR Exercise 4.

- While the gel is polymerizing, prepare the samples. Make two identical samples of 10 µg from the supernate from the homogenate of bovine heart and two identical samples of the 10 µg from the supernate from the homogenate from bovine skeletal muscle, as well as one sample from each of the other fractions. If a supernate of a homogenate of bovine skeletal muscle is not available, you will need to prepare two identical samples of each of the preparations you are going to run.

- Unfold the polypeptides in the various samples with dodecyl sulfate as described in FNR Exercise 4.

- Pour the stacking gel as described in FNR Exercise 4.

- Load the samples into the 10 lanes of the gel as suggested in one of the two following patterns:

- If your schedule included LDH Exercise 1, in which the homogenate of skeletal muscle was made, use the following distribution:

Lanes 1 and 5:	Commercial prestained markers.
Lane 2:	2 μg of supernate from the homogenate of skeletal muscle.
Lanes 3 and 10:	10 μg of supernate from the homogenate of skeletal muscle.
Lanes 4 and 9:	10 μg of supernate from the homogenate of heart.
Lane 6:	0.02 μg of purified LDH.
Lane 7:	0.1 μg of purified LDH.
Lane 8:	2 μg of supernate from the homogenate of heart.

  If this loading pattern is used, after electrotransfer the sheet of nitrocellulose will be cut between lanes 4 and 5.

- If your schedule did not include LDH Exercise 1, in which the homogenate of skeletal muscle was made, use the following distribution:

Lanes 1 and 6:	Commercial prestained markers.
Lanes 2 and 7:	0.02 μg of purified LDH.
Lanes 3 and 8:	0.1 μg of purified LDH.
Lanes 4 and 9:	2 μg of supernate from the homogenate of heart.
Lanes 5 and 10:	10 μg of supernate from the homogenate of heart.

  If this loading pattern is used, after electrotransfer the sheet of nitrocellulose will be cut between lanes 5 and 6.

- Perform the electrophoresis as described in FNR Exercise 4. While the electrophoresis is running, you should prepare everything you will need for the electrotransfer, as described in FNR Exercise 7. At the end of the run, record the distance migrated by the tracking dye. This will be needed for calculation of the R_f of each protein.

Electrotransfer to a sheet of nitrocellulose (20 min during first session followed by an overnight transfer).

- Set up and start the electrotransfer to the sheet of nitrocellulose as described in FNR Exercise 7. Electrotransfer the polypeptides in all 10 lanes of the gel onto the nitrocellulose. The electrotransfer is run overnight, so one of you will have to come in the next day to retrieve the sheet of nitrocellulose.

- The next day, while it is still in contact with the gel so that you can identify the lanes, cut the sheet of nitrocellulose with a fresh razor cleanly between lanes 4 and 5 or between lanes 5 and 6, depending on which loading pattern was used. The line of this cut is guided by the prestained markers that are visible in lane 5 or 6.

- Place the sheets of nitrocellulose carrying the proteins in the blocking solution of phosphate-buffered saline containing Tween. Use plastic trays slightly larger than the sheets.

If you are ready for the second session, proceed as described below. If the second session will be done on a later day, store the sheets of nitrocellulose carrying the proteins in the blocking solution at 4 °C.

SECOND SESSION

Immunodetection of the polypeptides from the A and B isoenzymatic subunits of LDH (4 h).

- Follow the instructions for immunodetection in FNR Exercise 7. In the present instance, however, distinct antibodies are used to detect the polypeptides of the A isoenzymatic subunit and the B isoenzymatic subunit of LDH, respectively.

If the samples of protein loaded on the gel were from heart and skeletal muscle:

- Lanes 1–4 are stained with the antibody for the A isoenzymatic polypeptide, which predominates in skeletal muscle, and lanes 5–10 are stained with the antibody for the B isoenzymatic polypeptide that predominates in the heart.

If the samples of protein loaded on the gel were only from heart:

- Lanes 1–5 are stained with the antibody for the A isoenzymatic polypeptide, which predominates in skeletal muscle, and lanes 6–10 are stained with the antibody for the B isoenzymatic polypeptide that predominates in the heart.

- In this exercise, the primary antibody should be diluted 1:1000 to give a final concentration of 1 μg mL^{-1}.

Analysis of the Data and Questions to Address in Your Report

1. Refer to question 1 in the Analysis of the Data and Questions to Address in Your Report of FNR Exercise 7 for the actual lengths and apparent lengths, in amino acids, of the prestained markers. Plot the negatives of the natural logarithms of the values for R_f of the markers as a function of their apparent lengths. Plot the negatives of the natural logarithms of the values for R_f of the markers as a function of their actual lengths.

Hint: The values of R_f will all be between 0 and 1. Thus, you will be plotting the negatives of natural logarithms of numbers between 0 and 1. In this case, a semilog plot generated by spreadsheet software with a logarithmic axis may be difficult to interpret. Thus it is best to calculate the negatives of the natural logarithms with a calculator or a spread sheet and then plot them on a linear scale.

Interpolate from the plot of the standards to estimate the length of the polypeptide detected by the anti-LDH antibodies. Is it possible to detect a protein multimer with this technique? Is the length you estimated the value expected for the length of a polypeptide of LDH from *B. taurus*?

2. The most critical property of an antiserum is specificity. Binding of an antibody to an epitope other than one in the antigen used to elicit the immune response in the host animal is *cross-reaction*. There are two types of cross-reaction: nonspecific and specific. If an antibody binds to a polypeptide unrelated in sequence to the expected target, *nonspecific* cross-reaction is occurring. This can be a serious limitation of immunostaining. *Specific* cross-reaction occurs when an antibody binds to a polypeptide similar enough in sequence to the antigen to have within its complete sequence of amino acids a part or parts identical or very close to the sequence of one or more epitopes in the sequence of the antigen. In the present instance, you will be depending on specific cross-reaction.

The primary antibody you are using to stain the A isoenzymatic polypeptide of bovine LDH was raised against a peptide with the sequence: ATLKDQLIYNLLKEEQTPQNKITVVGVGA VGMACAISILMKDLADELALV. The primary antibody you are using to stain the B isoenzymatic polypeptide of bovine LDH was raised against a peptide with the sequence: ENEVFLSLPCILNARGLTSVINQKLKDDEVAQLKKSADTLWDIQKDLKDL. In each case, antibodies in the antiserum that recognize epitopes in the peptide used as immunogen were then purified by affinity adsorption using a solid phase to which that peptide had been covalently attached. The former peptide is homologous to a sequence at the amino terminus of the A isoenzymatic polypeptide of bovine LDH. Align the sequence of the peptide with this sequence, which you can find at http://www.uniprot.org/uniprot/P19858. At the same time, align the homologous sequence from the amino terminus of the B isoenzymatic polypeptide, which you can find at http://www.uniprot.org/uniprot/Q5E9B1, with the alignment that you just made of the sequence of the peptide and the A isoenzymatic polypeptide. The latter peptide is homologous to a sequence at the carboxy terminus of the B isoenzymatic polypeptide of bovine LDH. Align the sequence of this peptide with this sequence of amino acids, which you can find at http://www.uniprot.org/uniprot/Q5E9B1. At the same time, align the homologous sequence of amino acids from the carboxy terminus of the A isoenzymatic polypeptide, which you can find at http://www.uniprot.org/uniprot/P19858, with the alignment that you just made of the sequence of the peptide and the sequence of the B isoenzymatic polypeptide. The easiest ways to perform these alignments is to cut and paste the relevant sequences. Use a Courier font to make these alignments. This is a font in which the width of each letter is the same. The alignments can be done easily by eye; there is no need to use an application.

3. Discuss what you observed in this experiment, the alignments of the sequences, and the problem of cross-reactivity.

4. Discuss the distribution of the isoenzymes of LDH among the various tissues that you observed in LDH Exercise 3 and the immunostaining you observed in this experiment. Include in your discussion a consideration of the large number of different proteins in the supernates of the homogenates.

5. In the technique of immunohistochemistry, an antigen, usually a particular protein, is detected in a frozen section of a particular tissue by immunostaining. Detergents are not typically used during preparation of the tissue. Suggest an explanation for the fact that some antibodies work well for immunoblots of polypeptides unfolded by dodecyl sulfate but not for immunohistochemistry, while the reverse is true for other antibodies.

6. Both the enzyme-linked immunosorbent assay (ELISA), mentioned in Technical Background in FNR Exercise 7, and the immunoblotting you just performed employ the chromogenic product of an enzymatic reaction for detection of the antigen. As explained in the two references provided in FNR Exercise 7, in the ELISA the colored product is soluble and

quantified by its absorbance. The method of ELISA is quantitative. Discuss whether or not the immunoblot you just performed is a quantitative method.

7. When the isoenzymes were submitted to electrophoresis, the medium used was cellulose acetate; when the polypeptides of LDH saturated with dodecyl sulfate were submitted to electrotransfer, the medium used to fix them in place was nitrocellulose. Both of these plastics are made from purified, natural cellulose. How do they differ in their properties that makes each one ideal for their respective purposes and each one useless for the other purpose?

Experimental Design

BACKGROUND

In each of the exercises in the previous sections, a detailed protocol and guidelines for analysis of data were provided. Such exercises allow you to learn the theory upon which the method is based and give you an opportunity to develop laboratory skills. Although this is a good way to prepare for participation in biochemical research, eventually you must be able to:

1. Create a hypothesis about a significant scientific question.

2. Design an experiment to test the hypothesis.

3. Conduct the experiment.

4. Evaluate the data to determine if the hypothesis is correct.

In each of the following exercises, a hypothesis and some technical hints about how to test it are provided. After completing the exercises from the earlier sections, you will be familiar with the equipment and supplies available in the lab. You will have used a variety of reagents and some of these reagents will be of use in the experiments you design for this section. The instructor should provide you with a list of other potentially useful reagents available in the lab.

Follow the guidelines in each of the following exercises to design and perform an appropriate experiment. Some suggestions are provided for analysis of the data you obtain. Because time is limited, these experiments are not at the level of or as complex as the research that you would encounter as a graduate student in biochemistry, but they should give you important practice working on your own.

SOURCES FOR BIOCHEMICAL METHODS

In addition to the present manual, there are a number of other biochemical laboratory manuals from various publishers that are also good technical sources (Ausubel, 1987, 2002; Boyer, 2000, 2012; Hardin, 2001; Jack, 1995; Marshak, 1996; Ninfa et al., 2010; Simpson et al., 2009; Switzer and Garrity, 1999; Williams et al., 2007). In addition to these books, there are collections of protocols online that are continually being updated. The examples listed below, however, require an institutional subscription for access.

Experiments in the Purification and Characterization of Enzymes.
http://dx.doi.org/10.1016/B978-0-12-409544-1.00004-7

Cold Spring Harbor Protocols: http://cshprotocols.cshlp.org.

Current Protocols from Wiley Online:
   Protein Science: http://onlinelibrary.wiley.com/book/10.1002/0471140864.
   Molecular Biology: http://onlinelibrary.wiley.com/book/10.1002/0471142727.
   Cell Biology: http://onlinelibrary.wiley.com/book/10.1002/0471143030.

EXPERIMENTAL DESIGN EXERCISE 1

Association of FNR with the Membranes of Thylakoids in Chloroplasts from Spinach

This exercise should only be attempted after FNR Exercises 1, 2, and 4 have been completed. Techniques used in those exercises will be needed for designing a proper experiment to address the hypothesis described below.

Hypothesis

The localization of ferredoxin-NADP$^+$ reductase (FNR) in chloroplasts from spinach was discussed earlier, in the section entitled Introduction to Enzymes Catalyzing Oxidation–Reductions with the Coenzyme NAD(P) and the Background for the experimental Section 1 covering FNR. There has been much evidence presented that at least some of the FNR molecules are dissolved in the aqueous phase of the stroma of the chloroplast. This phase is within the outer membrane of the chloroplast and surrounds the thylakoids. That FNR is free in solution in this space is consistent with the lack of a sequence of hydrophobic amino acids in the sequence for this protein (Figure 1.3) long enough to span a membrane and anchor the protein in one of the membranes in a chloroplast (Kyte, 2007, p. 764). Nevertheless, evidence for a specific association of FNR with membranes from chloroplasts has been presented (Grzyb et al., 2008; Lintala et al., 2007; Matthijs et al., 1986; Onda and Hase, 2004; Vallejos et al., 1984; Zhang et al., 2001). These reports, however, vary in the techniques employed and in their conclusions. Thus, it is worthwhile to test the hypothesis that FNR is associated with the thylakoid. If evidence for association of the enzyme with membranes from chloroplasts is obtained, other experiments could be designed to reveal the means by which FNR is linked to the thylakoid.

Technical Hints

Biological membranes are composed of phospholipids, sterols, and glycolipids (Kyte, 2007, pp 743–763). Detergents are a vital tool for assaying association of proteins with membranes (Kyte, 2007, pp 768–771). The exercises in which you submitted polypeptides saturated with dodecyl sulfate to electrophoresis are examples of the use of detergents in biochemical analysis. Dodecyl sulfate ion is a harsh detergent. While milder detergents dissolve membranes, they often do not unfold the polypeptides of proteins as dodecyl sulfate does. Enzymatic activity is also destroyed by dodecyl sulfate, so it is not possible to locate enzymes by their activity after treatment with dodecyl sulfate.

There are a variety of detergents used for characterizing the complexes between membranes and proteins (Table 4.1). These detergents, if used carefully, do not unfold polypeptides at concentrations at which they will dissolve the phospholipids and glycolipids that constitute biological

TABLE 4.1   Nondenaturing Detergents for the Study of Proteins Associated with Membranes[a]

Common Name	Systematic Name
Triton X-100	4-(1,1,3,3-Tetramethylbutyl)phenyl-polyethylene glycol
Nonidet P-40 (NP-40)	4-Nonylphenyl-polyethylene glycol
Cholate	3α,7α,12α-Trihydroxy-5β-cholan-24-oic acid
n-Octyl-D-glucoside	Octyl-β-D-glucopyranoside
CHAPS	3-[(3-Cholamidopropyl)dimethylammonio]-1-propanesulfonic acid

[a] *A more complete table listing the micellar properties of a number of nonionic and zwitterionic detergents has been published (Kyte, 2007).*

membranes. Some, but not all, are nonionic. In addition to the publications cited above, an earlier paper describing purification of plastoquinol–plastocyanin reductase (cytochrome b_6f) from the thylakoids of chloroplasts from spinach (Hurt and Hauska, 1981) may be of help.

Technical Preparation

Answer the following questions before designing your experiment:

1. How were the chloroplasts lysed in FNR Exercise 1?

2. What is the structure of the lipids in the thylakoids in a chloroplast?

3. What are the physical properties of the detergents in the table above that allow them to dissolve membranes and release the proteins from them?

4. Will any of these detergents disrupt the membrane of a thylakoid? If yes, explain why.

5. Is it possible that a detergent will release some but not all of the proteins associated with a membrane?

6. Which of the techniques used in the exercises already completed is appropriate for separating soluble proteins from those bound to membranes? What are the critical variables in the application of this technique?

7. Will the presence of an intact membrane or dispersed lipids affect an FNR activity assay?

Design the Experiment

Search databases and the Web sites of chemical companies for information about appropriate detergents. Look over Table 14.5 in the book cited earlier (Kyte, 2007). Some of the most commonly used databases are described in the Computational Techniques for Biochemistry section. For any reagent not already available for these exercises, the *Chem ACX* database is particularly useful because it gives supplier information.

Use what you have learned from the previous FNR exercises and the answers to the questions in the Technical Preparation section above to design the protocol. Electrophoresis of a protein sample derived from chloroplasts on a gel containing dodecyl sulfate that shows a band of the expected length cannot confirm the presence of FNR. Only a positive result from an assay for enzymatic activity or immunoelectrophoresis is conclusive evidence of the presence of the enzyme. Only enzymatic activity is quantitative.

Analyze the Data and Suggest Further Experiments

Evaluate both the results of this experiment that you have designed and the previous FNR exercises. Did the results of the previous FNR exercises provide evidence that FNR is localized within the chloroplast? If the results of this latest experiment provide evidence that FNR may be associated with the thylakoid, can you estimate the percentage of the molecules of enzyme in the chloroplast that are bound to the membrane? How can the way in which the protein is associated with the thylakoid be elucidated? How would you distinguish the binding of FNR to the membranes of the thylakoid from FNR free in solution but trapped in aqueous spaces within the thylakoids?

EXPERIMENTAL DESIGN EXERCISE 2

Search for Multiple Flavin Reductases in *Photobacterium leiognathi*

This exercise should only be performed after LuxG Exercises 1, 2, 4, and 5 have been completed. Techniques employed in those exercises will be critical for designing an experiment to address the hypothesis stated below. In addition, the data obtained from those exercises will be needed to interpret the new data.

Background

As mentioned earlier in the Background to the experimental Section 2, LuxG (Figure 2.1), the sequence of amino acids for Fre riboflavin reductase (NAD(P)H) of *Escherichia coli* is 36.4% identical to that of LuxG FMN reductase (NADH) from *P. leiognathi*. The former enzyme is the major flavin reductase of *E. coli* (Fieschi et al., 1995; Ingelman et al., 1999). In addition to the difference in the sequence of amino acids, LuxG and Fre may be distinguished by their specificity for substrates. Differences in the kinetic parameters are observed between the two enzymes when different combinations of the hydride donor (NADH or NADPH) and acceptor (flavin mononucleotide (FMN)), riboflavin, or flavin adenine dinucleotide (FAD) are used (Fieschi et al., 1995; Nijvipakul et al., 2008; Zenno and Saigo, 1994).

Hypothesis

Evidence for expression of both LuxG and Fre flavin reductases has been presented for some species of luminescent bacteria such as *Vibrio fischeri* and *Vibrio harveyi*. A coding sequence for what appears to be a Fre flavin reductase from *P. leiognathi* was posted in the UniProt database in March 2011 (http://www.uniprot.org). The sequence of amino acids encoded by this sequence of nucleotides is 44.9% identical to that of LuxG from this species. This entry, however, is derived from an entire sequence of the genome, and no experimental evidence has yet been presented that the encoded protein is actually a flavin reductase. Thus, it is a hypothesis that *P. leiognathi* expresses a Fre riboflavin reductase (NAD(P)H). The challenge is to find the most practical means of determining if this hypothesis is correct.

Technical Hints

The bacterial strain and growth media.
> Your instructor will provide the *P. leiognathi* strain. The optimal growth conditions for the various species in the genera *Photobacterium* and *Vibrio* are different from those for *E. coli*, as explained in a laboratory manual from investigators at the University of Wisconsin (Winfrey et al., 1997). This book provides the growth requirements for *V. fischeri*, which should work well enough for *P. leiognathi*.

Extraction of flavin reductases from P. leiognathi.
> The enzyme purifications in the earlier exercises make clear the procedural requirements for maintaining enzyme activity in cell lysates and during purification. The procedure in LuxG Exercise 1 provides some hints regarding lysis of *E. coli* bacteria.

Assaying for Fre activity in a lysate from P. leiognathi.
> The high degree of identity in sequence between LuxG and Fre suggests that it may be difficult to separate these two enzymes. Neither protein will have the histidine tag used for purifying recombinant LuxG-his$_6$ earlier. Thus, it may be best to make use of the difference in kinetic properties mentioned above. The procedures in LuxG Exercise 1 and

LDH Exercises 1 and 2 provide some guidance regarding assaying enzymatic activity in an unfractionated lysate. Refer to Introduction to Enzymes Catalyzing Oxidation–Reductions with the Coenzyme NAD(P) regarding the catalytic constant k_0 and the Michaelis constant K_m.

Technical Preparation

Try to answer the following questions before designing your experiment:

1. Refer to the references mentioned above to obtain estimates of the values of K_m for the various reactants with the LuxG and Fre enzymes. What reaction conditions are necessary to obtain initial velocities (v) that will be directly proportional to the catalytic constant k_0 for a particular enzyme–reactant combination?

2. How can the differences in the values of K_m of LuxG and Fre for different reactants, discussed in the references given above, be used to determine if *P. leiognathi* expresses both reductases or only LuxG?

3. To measure the K_m of an enzyme for a particular substrate, what factors determine the choice of concentrations of reactant that will be used?

Design the Experiment

Use the preceding discussion and your answers to the questions in the Technical Preparation section above to plan the experiment. Decide which of the reactant combinations is best for the assay. Consider the possible outcomes of this assay. What conclusion may be drawn from each possible outcome?

Analyze the Data and Suggest Further Experiments

Compare the data obtained from your measurements to that in the references cited above. When interpreting experimental results it is important to distinguish between data *consistent* with a hypothesis and data that *prove* a hypothesis is true. Answer the following:

1. Is the relative abundance of enzymes in *P. leiognathi* an issue? Why or why not?

2. Are the data obtained consistent with the expression of Fre in *P. leiognathi*?

3. If the answer to question 2 is yes, do the data prove that this bacterium expresses Fre? If the answer to question 2 is no, do the data prove that this bacterium does not express Fre?

Suggest experiments that would make your conclusion more definite. How could Fre and LuxG be separated? Find an estimate of the isoionic points of the two proteins and discuss the separation of the two native proteins by electrophoresis. Suggest experiments to investigate whether *P. leiognathi* expresses a flavin reductase distinct from either Fre or LuxG.

EXPERIMENTAL DESIGN EXERCISE 3

Assay for LDH Activity Specific to (R)-lactate in Bovine Mitochondria

D-Lactate dehydrogenase (cytochrome) (EC 1.1.2.4; D-LDH) uses (R)-lactate as the reductant and oxidized cytochrome c as the oxidant (Figure 4.1).

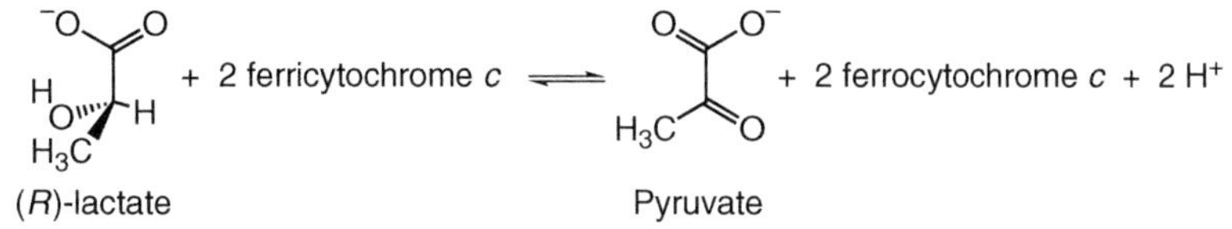

FIGURE 4.1   Reaction catalyzed by D-lactate dehydrogenase (cytochrome).

It is best if you have performed exercises from the experimental Section 3, LDH, involving preparation of extracts from bovine tissue and assays for the enzymatic activity of L-LDH before attempting to design an experiment studying D-LDH. Although not essential, it will also be helpful if you have completed FNR Exercise 1 (the isolation of chloroplasts) and LuxG Exercise 1 (the lysis of bacteria by sonication); if you have not, you should read these sections.

Background

As explained earlier in the Background to the LDH experimental section, the sequences of amino acids for L-LDH and D-LDH have no similarity (Engqvist et al., 2009). This suggests that they arose independently during evolution. The fact that the two enzymes are present in distinct cellular compartments in eukaryotes, L-LDH in the cytoplasm and D-LDH in the mitochondrion, is also consistent with independent origin. Most of the study of D-LDH has been in species other than animals. Some experiments on recombinant mammalian D-LDH (Flick and Konieczny, 2002) and the endogenous enzyme in rat liver (de Bari et al., 2002) have been published.

2-Oxopropanal (methylglyoxal; Figure 4.2) is a toxic molecule that is a side product from the interconversion of dihdroxyacetone phosphate and glyceraldehyde-3-phosphate, catalyzed by triose-phosphate isomerase in the normal course of glycolysis. In eukaryotes, the enzyme that

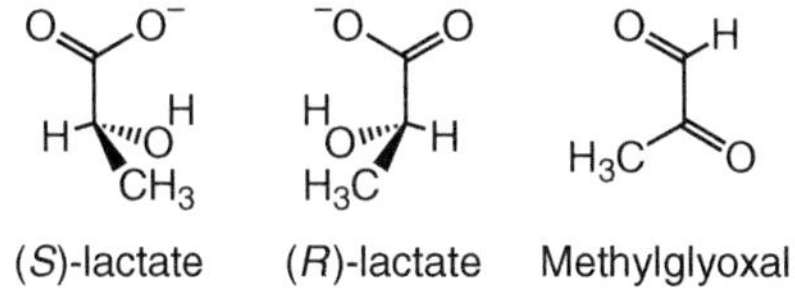

FIGURE 4.2   **Structures of the two stereoisomers of lactate and the structure of methylglyoxal.**   The Cahn–Ingold–Prelog (R,S) system for naming stereoisomers is recommended by the International Union of Biochemistry and Molecular Biology and is used here. The Fischer–Rosanoff (D,L) system, however, is also accepted for naming such pairs of structures. For the two forms of lactate, S corresponds to L and R corresponds to D.

protects the cell from the toxic effects of 2-oxopropanal is lactoylglutathione lyase. Cytoplasmic 2-oxopropanal is converted by lactoylglutathione lyase to (R)-S-lactoylglutathione, which then hydrolyses spontaneously to (R)-lactate. The (R)-lactate is transported into the mitochondrion (de Bari et al., 2002; Ewaschuk et al., 2005). The predominant theory at the moment is that D-LDH converts (R)-lactate to pyruvate, and the pyruvate then enters the citric acid cycle.

The electron acceptor coupled with D-lactate oxidation is cytochrome *c* (Figures 1.23 and 4.1), which is enclosed within the inner mitochondrial membrane (de Bari et al., 2002; Engqvist et al., 2009). This mechanism is believed to both remove a toxic molecule and capture the energy in the 2-oxopropanal.

When this system is functioning properly in a healthy mammal, the concentration of (*R*)-lactate in serum is in the nanomolar range (Ewaschuk et al., 2005). There are circumstances, however, in which the concentration reaches the millimolar range. When this occurs, (*R*)-lactate itself becomes toxic.

The primary substrates for the enzymes D-LDH and 2-hydroxyglutarate dehydrogenase are (*S*)-2-hydroxy acids. Although each enzyme is named for the substrate that is best catalyzed, some cross-reaction with other substrates has been documented for these enzymes (Achouri et al., 2004; Engqvist et al., 2009). It is critical to consider this when attempting to identify the enzymes present in an extract from a cell or an organelle.

Hypothesis

The mitochondrion plays a fundamental role in eukaryotic cellular processes. In all eukaryotic organisms in which it has been characterized, D-lactate dehydrogenase (cytochrome) has been found to be located within the mitochondria. The only mammalian tissue, however, examined to date for D-LDH activity is liver from the rat. The hypothesis is that the role of D-LDH in detoxification is fundamental and that D-LDH is present in all mammalian cell types. This may be tested by fractionating cells from mammalian tissues other than liver from the rat.

Technical Hints

Processing of tissue and fractionation of cells.

Purification to homogeneity is necessary for a complete characterization of an enzyme, but experiments with only partially purified enzymes are a good initial attempt to address the hypothesis stated above. The LDH exercises you have already performed that include cellular lysis and fractionation provide some clues regarding separating mitochondria from the remainder of a mammalian cell. Chloroplasts have some similarity to mitochondria. Thus, FNR Exercise 1 also provides some hints.

Assay of enzymatic activity.

Detection of an enzymatic activity in a subcellular compartment does not prove that the enzyme is *specific* to that compartment. To prove specificity, it is necessary to compare the concentration of enzymatic activity in various subcellular fractions. In addition, consider whether it will be worthwhile to assay for activities other than D-LDH.

Sources for protocols in the scientific literature.

The intense interest in the contents and processes within the mitochondrion has generated many book chapters and journal articles describing manipulation of this organelle. Some particularly relevant examples are in the bibliography (Boyer, 2000; de Bari et al., 2002; de Duve et al., 1955). The sources for biochemical methods provided at the beginning of this section should also be consulted. It is critical, however, to be aware of the differences between what you are trying to accomplish and what the authors of the cited articles were accomplishing with their methods.

Technical Preparation

Try to answer the following questions before designing the experiment. The articles cited above should be helpful.

1. What aspects of the structure of the mitochondrion must be considered in planning a method to assay enzyme activity localized in this organelle?

2. What is the best method for lysing bovine cells without disrupting the mitochondria?

3. Consider the equipment, supplies, and reagents available to you. In addition to the mitochondrial fraction, which other subcellular fractions will you be able to isolate and assay?

4. What will be the most practical assay for mammalian D-lactate dehydrogenase (cytochrome)?

5. How will you distinguish between D-LDH activity and the activity of other 2-hydroxy acid dehydrogenases?

6. Consider the options for lysing a subcellular organelle. How may mitochondria be lysed without inactivating the enzymes that are within them?

Design the Experiment

Use the technical hints and your answers to the questions in the Technical Preparation section above to design an experiment that will address the hypothesis. Bovine tissues are the easiest to obtain in large quantity. It would be most interesting to use a tissue other than liver because that is the one mammalian tissue in which D-LDH has already been detected. Mitochondrial function depends on the availability of O_2. Of the bovine tissues used in exercises in Section 3, the LDH experimental section, which has the greatest access to O_2? Consult the cell biology and biochemistry literature to find out if this tissue has more mitochondria in each cell than the other tissues.

Analyze the Data and Suggest Further Experiments

When evaluating results of experiments, it is important to distinguish between evidence that is *consistent* with a hypothesis and evidence that *proves* that a hypothesis is correct. Are the data from this experiment consistent with the hypothesis? Is there a practical way to prove, beyond all doubt, that the hypothesis is correct? The articles cited above explain the current knowledge regarding conversion of methylglyoxal to (R)-lactate. Do the data obtained from the experiment address this mechanism?

What circumstances may lead to a rise in the concentration of (R)-lactate in the serum of a mammal? The concentration of (S)-lactate in the serum of a healthy mammal is in the millimolar range. Why is the R isomer toxic at this concentration but not the L isomer? To understand the difference in the tolerance of mammals to the two lactate isomers, it may be useful to determine the relative amounts of the L-LDH and D-LDH enzymes in each cell in various tissues. Can you design an experiment that would address this issue?

Appendix I

MEASUREMENT OF ABSORBANCE WITH MULTIWELL PLATES AND AN AUTOMATED PLATE READER

An alternative to using a conventional spectrophotometer for assaying the absorption of ultraviolet or visible light is a plate reader. Plate readers (Figure 1.13) offer the following advantages:

1. A smaller volume of each sample (0.2 mL) is used. Conventional spectrophotometry typically requires between 0.5 and 3 mL for each measurement. Thus, less of a precious protein sample need be sacrificed for the measurement. This also conserves staining reagents, such as that used for the Bradford assay.

2. It is possible to assay 96 samples consecutively in a single multiwell plate. If large numbers of samples need to be assayed, storage and handling are more convenient with multiwell plates than with disposable test tubes or cuvettes that hold only one sample each.

3. When a single multiwell plate is inserted into the plate reader, as many as 96 samples may be assayed. This is much easier than inserting 96 tubes or cuvettes, one at a time, into a conventional spectrophotometer.

CHOOSING THE APPROPRIATE TYPE OF PLATE

Multiwell plates (Figure 1.13) are made from several different types of plastic. For absorbance of visible light, that with a wavelength greater than 400 nm, standard plastic works best. For wavelengths less than 400 nm, however, a plastic that is transparent to ultraviolet light is used. Plastic transparent to ultraviolet light is not completely transparent to visible light. Wells must have a flat bottom for consistent readings. The following are suggested brands:

Nunc 269620: standard plastic, 96 flat-bottom wells.

Costar 3635: transparent to ultraviolet light, 96 flat-bottom wells.

The plates that are transparent to ultraviolet light are more expensive than the plates made of standard plastic so it is best to reuse them. Rinse them with distilled water before the samples dry in the wells.

PREPARING SAMPLES OF PROTEIN AND OPERATING THE PLATE READER

There will probably be only one machine and many students. Therefore, it may be best if your instructor or teaching assistant operates the plate reader. Each plate to be assayed should have at least one well containing 0.2 mL of the appropriate blank solution for the experiment: staining reagent with no protein for the Bradford assay, a negative control reaction mixture for activity assays, or buffered solution lacking protein for the A_{280} measurement.

In some cases you may be using a plate layout with a designated well containing a blank sample or negative control. If such a layout is used, the sample for the blank or the negative control must be in the appropriate well on each plate. The computer program used for interpretation of the data will subtract the signal detected in the blank or negative control well from all the other readings. Since each plate has 96 wells, samples from several students (or pairs of students if working as partners) may be loaded in each plate.

Appendix II

OPERATION OF THE SPECTRONIC 20D+ SPECTROPHOTOMETER

This device is typically referred to as a *Spec 20*. Wavelengths from 340 to 625 nm may be assayed. Disposable glass test tubes are typically used to hold the samples. The minimum volume of sample needed is 2.5 mL, however 3.0 mL is usually used. Light of wavelength 340 to 400 nm is technically in the ultraviolet range, and the glass test tubes are not completely transparent to such light. A large enough portion of the signal in this range does pass through the glass, however, for the device to be useful for assays employing these wavelengths.

Although you will be using it for measurements of absorbance, it is best to set the baseline initially in transmittance mode. After achieving a 100% transmittance reading with a blank (water or the negative control sample for the assay), switch to absorbance mode. The reading should be close to zero.

The display reading returns to 0% transmittance whenever the tube containing the sample is removed from the sample holder, because removing the tube releases an occluder, which is an opaque piece of plastic that drops into the light beam and prevents the beam from reaching the phototube. Adjusting the zero transmittance on the scale can therefore always be done by simply removing the tube.

1. *Set the baseline in transmittance mode.*

The spectrophotometer is turned on by rotating the amplifier control (left-hand knob) clockwise. This should be done at least 20 min before measurements are made. Set the mode switch to "transmittance" and select the appropriate wavelength. After the instrument has warmed up, adjust the amplifier control knob so that the display reads 0% transmittance when no light is striking the phototube (no tube in holder).

The right-hand (light-control) knob regulates the amount of light passing through the second slit to the phototube. This light-control knob is needed because the light source emits light of different wavelengths at different intensities and the phototube is not equally responsive to light of different wavelengths. In addition, the blank solution used to zero the instrument, usually the solvent in which the light-absorbing species, or *chromophore*, is dissolved, may itself absorb light of certain wavelengths. In order to measure the absorbance due to only the chromophore, the intensity of the light is adjusted to compensate for this background absorbance.

After the spectrophotometer has been zeroed when no light is hitting the photocell, a blank solution is placed in the light path and the light-control knob is rotated until the display reads 100% transmittance. This adjustment achieves the desired compensation. If a sample solution is now placed in the light path, any change in the percent transmittance is due only to the particular chromophore in the sample and is related to the concentration of that chromophore. Whenever a change in wavelength is made, the 0%

and 100% transmittance must be reset, since the amount of compensation needed varies with wavelength.

2. *Switch to the absorbance mode and collect the data.*
   With the blank sample still in the holder, switch to absorbance mode. The reading on the display should now be close to zero. Bring it to exactly zero with the light-control knob. You may now start assaying experimental samples.

PROPERTIES OF THE PHOTOTUBE

The phototube is a photoemissive cell with a surface of the cesium–antimony type. The relative response of the phototube to a beam of monochromatic light of constant intensity is summarized in Table a.1.

TABLE a.1   Properties of the Phototube in a Spectronic 20D+ Spectrophotometer

Wavelength	Response Relative to Response at 400 nm
350	0.90
400	1.0
450	0.91
500	0.68
550	0.37
600	0.10
625	0.05

It can be seen that the phototube is much more sensitive to light of wavelength 400 nm than to light of wavelength 600 nm. This means that the phototube will require a greater influx of monochromatic light at 600 than at 400 nm in order for the same transmittance reading to be registered upon the colorimeter dial. This is why the adjustment has to be made to 100% transmittance whenever the wavelength is changed.

HANDLING THE TUBES FOR SAMPLES WITH THE SPECTRONIC 20D+

The handling of the tubes is extremely important. Any variation in the design of the tubes, such as variation in the width or the curvature of the glass, may create undesired variation in the data obtained. Stains, smudges, or scratches may also cause problems. The following precautions are advised:

1. Be certain that all tubes used for a particular experiment are of the same brand and design.
2. Do not touch the lower portion of the tubes through which the light beam will pass.
3. Wipe off any liquid drops or smudges on the lower half of the tube with a clean Kimwipe before placing it in the instrument. Never wipe the tubes with towels or handkerchiefs.
4. If the tube has been in the cold room or on ice, allow it to warm to room temperature before inserting it into the Spec 20, or condensation may form on the glass while you are taking your measurement. Check for condensation when you remove the tube from the spectrophotometer.

Bibliography

Achouri, Y., Noël, G., Vertommen, D., et al., 2004. Identification of a dehydrogenase acting on D-2-hydroxyglutarate. Biochem. J. 381, 35–42.

Ackers, G.K., 1964. Molecular exclusion and restricted diffusion processes in molecular-sieve chromatography. Biochemistry. 3, 723–730.

Aliverti, A., Jansen, T., Zanetti, G., et al., 1990. Expression in *Escherichia coli* of ferredoxin:NADP+ reductase from spinach: Bacterial synthesis of the holoflavoprotein and of an active enzyme form lacking the first 28 amino acid residues of the sequence. Eur. J. Biochem. 191, 551–555.

Andrews, P., 1965. The gel-filtration behaviour of proteins related to their molecular weights over a wide range. Biochem. J. 96, 595–606.

Apley, E.C., Wagner, R., Engelbrecht, S., 1985. Rapid procedure for the preparation of ferredoxin-NADP+ oxidoreductase in molecularly pure form at 36 kDa. Anal. Biochem. 150, 145–154.

Arakawa, T., Timasheff, S.N., 1984. Mechanism of protein salting in and salting out by divalent cation salts: Balance between hydration and salt binding. Biochemistry. 23, 5912–5923.

Ausubel, F.M., 1987. Current Protocols in Molecular Biology. Greene Publishing Associates/Wiley, New York.

Ausubel, F.M., 2002. Short Protocols in Molecular Biology: A Compendium of Methods from Current Protocols in Molecular Biology, fifth ed. Wiley, New York.

Aviva Systems Biology, 2012. http://www.avivasysbio.com.

Barber, J., 2009. Photosynthetic energy conversion: Natural and artificial. Chem. Soc. Rev. 38, 185–196.

Barnett, H., 1964. The staining of lactic dehydrogenase isoenzymes after electrophoretic separation on cellulose acetate. J. Clin. Pathol. 17, 567–570.

Barton, J.S., 2011. A comprehensive enzyme kinetic exercise for biochemistry. J. Chem. Educ. 88, 1336–1339.

Berg, J.M., Tymoczko, J.L., Stryer, L., 2012. Biochemistry, seventh ed. W. H. Freeman, New York.

Bergmeyer, H.U., 1975. New values for molar extinction coefficients of NADH and NADPH for use in routine laboratories. Z. Klin. Chem. Klin. Bio. 13, 507–508.

Binda, C., Coda, A., Aliverti, A., et al., 1998. Structure of the mutant E92K of [2Fe-2S] ferredoxin I from *Spinacia oleracea* at 1.7 A resolution. Acta. Crystallogr. D. Biol. Crystallogr. 54, 1353–1358.

Binder, A., Jagendorf, A., Ngo, E., 1978. Isolation and composition of the subunits of spinach chloroplast coupling factor protein. J. Biol. Chem. 253, 3094–3100.

Boyer, R.F., 2000. Modern Experimental Biochemistry, third ed. Benjamin Cummings, San Francisco, CA.

Boyer, R.F., 2012. Biochemistry Laboratory: Modern Theory and Techniques, second ed. Prentice Hall, Boston, MA.

Bradford, M., 1976. A rapid and sensitive method for the quantitation of microgram quantities of protein utilizing the principle of protein-dye binding. Anal. Biochem. 72, 248–254.

Bruns, C.M., Karplus, P.A., 1995. Refined crystal structure of spinach ferredoxin reductase at 1.7 A resolution: Oxidized, reduced and 2'-phospho-5'-AMP bound states. J. Mol. Biol. 247, 125–145.

Burnette, W.N., 1981. "Western blotting": Electrophoretic transfer of proteins from sodium dodecyl sulfate–polyacrylamide gels to unmodified nitrocellulose and radiographic detection with antibody and radioiodinated protein. A. Anal. Biochem. 112, 195–203.

Cacace, M.G., Landau, E.M., Ramsden, J.J., 1997. The Hofmeister series: Salt and solvent effects on interfacial phenomena. Q Rev. Biophys. 30, 241–277.

Chaikuad, A., Fairweather, V., Conners, R., et al., 2005. Structure of lactate dehydrogenase from *Plasmodium vivax*: Complexes with NADH and APADH. Biochemistry 44, 16221–16228.

Chakhtoura, M., Abdelnoor, A.M., 2010. Monoclonal antibodies used as prophylactic, therapeutic and diagnostic agents. Immunopharmacol. Immunotoxicol. 32, 533–542.

Champagne, C.D., Houser, D.S., Crocker, D.E., 2005. Glucose production and substrate cycle activity in a fasting adapted animal, the northern elephant seal. J. Exp. Biol. 208, 859–868.

Cross, A.R., Yarchover, J.L., Curnutte, J.T., 1994. The superoxide-generating system of human neutrophils possesses a novel diaphorase activity: Evidence for distinct regulation of electron flow within NADPH oxidase by p67-phox and p47-phox. J. Biol. Chem. 269, 21448–21454.

Cuatrecasas, P., 1970a. Agarose derivatives for purification of protein by affinity chromatography. Nature 228, 1327–1328.

Cuatrecasas, P., 1970b. Protein purification by affinity chromatography: Derivatizations of agarose and polyacrylamide beads. J. Biol. Chem. 245, 3059–3065.

Dale, J., Schantz, M.V., Plant, N., 2011. From Genes to Genomes: Concepts and Applications of DNA Technology, third ed. John Wiley & Sons, Chichester, UK.

Davis, B., 1964. Disc electrophoresis II: Method and application to human serum proteins. Ann. NY Acad. Sci. 121, 404–427.

de Bari, L., Atlante, A., Guaragnella, N., et al., 2002. D-lactate transport and metabolism in rat liver mitochondria. Biochem. J. 365, 391–403.

de Duve, C., Pressman, B.C., Gianetto, R., et al., 1955. Tissue fractionation studies 6: Intracellular distribution patterns of enzymes in rat-liver tissue. Biochem. J. 60, 604–617.

Dietz, A.A., Lubrano, T., 1967. Separation and quantitation of lactic dehydrogenase isoenzymes by disc electrophoresis. Anal. Biochem. 20, 246–257.

Edelhoch, H., 1967. Spectroscopic determination of tryptophan and tyrosine in proteins. Biochemistry 6, 1948–1954.

Elia, G., 2010. Protein biotinylation. Curr. Protoc. Protein. Sci. Chapter 3, Unit 3.6.

Engqvist, M., Drincovich, M.F., Flügge, U.I., Maurino, V.G., 2009. Two D-2-hydroxy-acid dehydrogenases in *Arabidopsis thaliana* with catalytic capacities to participate in the last reactions of the methylglyoxal and beta-oxidation pathways. J. Biol. Chem. 284, 25026–25037.

Engvall, E., 1980. Enzyme immunoassay ELISA and EMIT. Methods. Enzymol. 70, 419–439.

Eszes, C.M., Sessions, R.B., Clarke, A.R., et al., 1996. Removal of substrate inhibition in a lactate dehydrogenase from human muscle by a single residue change. FEBS Lett. 399, 193–197.

Everse, J., Kaplan, N.O., 1973. Lactate dehydrogenases: Structure and function. Adv. Enzymol. Relat. Areas. Mol. Biol. 37, 61–133.

Everse, J., Zoll, E., Kahan, L., Kaplan, N.O., 1971. Addition products of diphosphopyridine nucleotides with substrates of pyridine-nucleotide-linked dehydrogenases. Bioorgan. Chem. 1, 207–233.

Ewaschuk, J.B., Naylor, J.M., Zello, G.A., 2005. D-lactate in human and ruminant metabolism. J. Nutr. 135, 1619–1625.

Fieschi, F., Niviere, V., Frier, C., et al., 1995. The mechanism and substrate specificity of the NADPH:flavin oxidoreductase from *Escherichia coli*. J. Biol. Chem. 270, 30392–30400.

Flick, M.J., Konieczny, S.F., 2002. Identification of putative mammalian D-lactate dehydrogenase enzymes. Biochem. Biophys. Res. Commun. 295, 910–916.

Gershoni, J.M., Palade, G.E., 1983. Protein blotting: Principles and applications. Anal. Biochem. 131, 1–15.

Goldberg, E., Eddy, E.M., Duan, C., Odet, F., 2010. LDHC: The ultimate testis-specific gene. J. Androl. 31, 86–94.

Grzyb, J., Gagos, M., Gruszecki, W.I., et al., 2008. Interaction of ferredoxin:NADP+ oxidoreductase with model membranes. Biochim. Biophys. Acta. 1778, 133–142.

Gulotta, M., Deng, H., Dyer, R.B., Callender, R.H., 2002. Toward an understanding of the role of dynamics on enzymatic catalysis in lactate dehydrogenase. Biochemistry 41, 3353–3363.

Gutierrez, A., Grunau, A., Paine, M., et al., 2003. Electron transfer in human cytochrome P450 reductase. Biochem. Soc. Trans. 31, 497–501.

Hardin, C., 2001. Cloning, Gene Expression, and Protein Purification: Experimental procedures and process rationale. Oxford University Press, New York.

Harlow, E., Lane, D., 1999. Using Antibodies: A laboratory manual. Cold Spring Harbor Laboratory Press, Cold Spring Harbor, NY.

Holbrook, J.J., 1975. In: Boyer, P.D. (Ed.), The Enzymes. Academic Press, New York, pp. 191–292.

Hondred, D., Hanson, A.D., 1990. Hypoxically inducible barley lactate dehydrogenase: cDNA cloning and molecular analysis. Proc. Natl. Acad. Sci. USA. 87, 7300–7304.

Hosseinkhani, S., Szittner, R., Meighen, E., 2005. Random mutagenesis of bacterial luciferase: Critical role of Glu175 in the control of luminescence decay. Biochem. J. 385, 575–580.

Hurt, E., Hauska, G., 1981. A cytochrome f/b6 complex of five polypeptides with plastoquinol-plastocyanin-oxidoreductase activity from spinach chloroplasts. Eur. J. Biochem. 117, 591–595.

Ingelman, M., Ramaswamy, S., Nivière, V., et al., 1999. Crystal structure of NAD(P)H:flavin oxidoreductase from *Escherichia coli*. Biochemistry 38, 7040–7049.

Jack, R.C., 1995. Basic biochemical laboratory procedures and computing: With principles, review questions, worked examples, and spreadsheet solutions. Oxford University Press, New York.

Kaplan, N.O., Everse, J., Dixon, J.E., et al., 1974. Purification and separation of pyridine nucleotide-linked dehydrogenases by affinity chromatography techniques. Proc. Natl. Acad. Sci. USA. 71, 3450–3454.

Karplus, P.A., Daniels, M.J., Herriott, J.R., 1991. Atomic structure of ferredoxin-NADP+ reductase: Prototype for a structurally novel flavoenzyme family. Science 251, 60–66.

Kemmer, G., Keller, S., 2010. Nonlinear least-squares data fitting in Excel spreadsheets. Nat. Protoc. 5, 267–281.

Kietzmann, T., 2004. Oxygen-dependent regulation of hepatic glucose metabolism. Methods. Enzymol. 381, 357–376.

Kohler, G., Milstein, C., 1976. Derivation of specific antibody-producing tissue culture and tumor lines by cell fusion. Eur. J. Immunol. 6, 511–519.

Kozmin, S., Wang, J., Schaaper, R., 2010. Role for CysJ flavin reductase in molybdenum cofactor-dependent resistance of *Escherichia coli* to 6-N-hydroxylaminopurine. J. Bacteriol. 192, 2026–2033.

Krebs, J.E., Lewin, B., Kilpatrick, S.T., Goldstein, E.S., 2013. Lewin's Genes XI, eleventh ed. Jones & Bartlett Learning, Burlington, MA.

Kurien, B.T., Scofield, R.H., 2009. Introduction to protein blotting. Methods. Mol. Biol. 536, 9–22.

Kyte, J., 1995. Mechanism in Protein Chemistry. Garland, New York.

Kyte, J., 2007. Structure in Protein Chemistry, second ed. Garland Science, New York.

Laemmli, U., 1970. Cleavage of structural proteins during the assembly of the head of bacteriophage T4. Nature 227, 680–685.

Lehninger, A.L., Nelson, D.L., Cox, M.M., 2008. Lehninger Principles of Biochemistry, fifth ed. W.H. Freeman, New York.

Lewis, M., Chang, G., Horton, N.C., et al., 1996. Crystal structure of the lactose operon repressor and its complexes with DNA and inducer. Science 271, 1247–1254.

Lintala, M., Allahverdiyeva, Y., Kidron, H., et al., 2007. Structural and functional characterization of ferredoxin-NADP+-oxidoreductase using knock-out mutants of *Arabidopsis*. Plant. J. 49, 1041–1052.

Mager, M., Blatt, W.F., Abelmann, W.H., 1966. The use of cellulose acetate for the electrophoretic separation and quantitation of serum lactic dehydrogenase isozymes in normal and pathologic states. Clin. Chim. Acta. 14, 689–697.

Malmstrom, J., Lee, H., Nesvizhskii, A.I., et al., 2006. Optimized peptide separation and identification for mass-spectrometry-based proteomics via free-flow electrophoresis. J. Proteome. Res. 5, 2241–2249.

Margoliash, E., Frohwirt, N., 1959. Spectrum of horse-heart cytochrome c. Biochem. J. 71, 570–572.

Marshak, D.R., 1996. Strategies for Protein Purification and Characterization: A laboratory course manual. Cold Spring Harbor Laboratory Press, Plainview, NY.

Matthijs, H.C., Coughlan, S.J., Hind, G., 1986. Removal of ferredoxin:NADP+ oxidoreductase from thylakoid membranes, rebinding to depleted membranes, and identification of the binding site. J. Biol. Chem. 261, 12154–12158.

Mayhew, S.G., 1999. The effects of pH and semiquinone formation on the oxidation-reduction potentials of flavin mononucleotide: A reappraisal. Eur. J. Biochem. 265, 698–702.

McComb, R.B., Bond, L.W., Burnett, R.W., et al., 1976. Determination of the molar absorptivity of NADH. Clin. Chem. 22, 141–150.

Michaelis, L., Menten, M.L., 1913. The kinetics of the inversion effect. Biochem. Z. 49, 333–369.

Michaelis, L., Menten, M.L., Johnson, K.A., Goody, R.S., 2011. The original Michaelis constant: Translation of the 1913 Michaelis–Menten paper. Biochemistry 50, 8264–8269.

Miller, W., Huang, N., Agrawal, V., Giacomini, K., 2009. Genetic variation in human P450 oxidoreductase. Mol. Cell. Endocrinol. 300, 180–184.

Moczydlowski, E.G., Fortes, P.A., 1981. Characterization of 2′,3′-O-(2,4,6-trinitrocyclohexadienylidine)adenosine 5′-triphosphate as a fluorescent probe of the ATP site of sodium and potassium transport adenosine triphosphatase: Determination of nucleotide binding stoichiometry and ion-induced changes in affinity for ATP. J. Biol. Chem. 256, 2346–2356.

Monthony, J., Wallace, E., Allen, D., 1978. A non-barbital buffer for immunoelectrophoresis and zone electrophoresis in agarose gels. Clin. Chem. 24, 1825–1827.

Mosbach, K., Guilford, H., Ohlsson, R., Scott, M., 1972. General ligands in affinity chromatography: Cofactor-substrate elution of enzymes bound to the immobilized nucleotides adenosine 5′-monophosphate and nicotinamide-adenine dinucleotide. Biochem. J. 127, 625–631.

Negi, S.S., Carol, A.A., Pandya, S., et al., 2008. Co-localization of glyceraldehyde-3-phosphate dehydrogenase with ferredoxin-NADP reductase in pea leaf chloroplasts. J. Struct. Biol. 161, 18–30.

Nicholas, K.B., Nicholas, H.B., Deerfield, D.W., 1997. GeneDoc: Analysis and visualization of genetic variation. Embnetnews 4, 1–4.

Nijvipakul, S., Wongratana, J., Suadee, C., et al., 2008. LuxG is a functioning flavin reductase for bacterial luminescence. J. Bacteriol. 190, 1531–1538.

Ninfa, A.J., Ballou, D.P., Benore, M., 2010. Fundamental Laboratory Approaches for Biochemistry and Biotechnology, second ed. John Wiley & Sons, Hoboken, NJ.

Novagen, 1998. pET-24a-d+ vectors. 070 , Tech. Bull. http://www.emdmillipore.com/life-science-research/technical-bulletins.

Novagen, 2006a. His-bind kits. 054 , Tech. Bull. http://www.emdmillipore.com/life-science-research/technical-bulletins.

Novagen, 2006b. pET system manual. 055 , Tech. Bull. http://www.emdmillipore.com/life-science-research/-technical-bulletins.

Novagen, 2007. Protein purification and detection tools, 2nd ed. http://www.emdmillipore.com/life-science-research/technical-bulletins.

Novagen, 2012. Novagen. Technical. Bulletins.. http://www.emdmillipore.com/life-science-research/technical-bulletins.

Onda, Y., Hase, T., 2004. FAD assembly and thylakoid membrane binding of ferredoxin:NADP+ oxidoreductase in chloroplasts. FEBS Lett. 564, 116–120.

Opher, A.W., Collier, C.S., Miller, J.M., 1966. A rapid electrophoretic method for the determination of the isoenzymes of serum lactate dehydrogenase. Clin. Chem. 12, 308–313.

Ornstein, L., 1964. Disc electrophoresis. I. Background and theory. Ann. NY. Acad. Sci. 121, 321–349.

Pace, C., Vajdos, F., Fee, L., et al., 1995. How to measure and predict the molar absorption coefficient of a protein. Protein. Sci. 4, 2411–2423.

Perkins, S.J., 1986. Protein volumes and hydration effects: The calculations of partial specific volumes, neutron scattering matchpoints and 280-nm absorption coefficients for proteins and glycoproteins from amino acid sequences. Eur. J. Biochem. 157, 169–180.

Pringle, J.R., 1970. The molecular weight of the undegraded polypeptide chain of yeast hexokinase. Biochem. Biophys. Res. Commun. 39, 46–52.

Pugliese, L., Coda, A., Malcovati, M., Bolognesi, M., 1993. Three-dimensional structure of the tetragonal crystal form of egg-white avidin in its functional complex with biotin at 2.7 A resolution. J. Mol. Biol. 231, 698–710.

Rahman, G., Lim, J.Y., Jung, K.D., Joo, O.S., 2010. NAD(+) hydrogenation on Au electrode deposited on modified glassy carbon. Electrochem. Commun. 12, 1371–1374.

Read, J., Winter, V., Eszes, C., et al., 2001. Structural basis for altered activity of M- and H-isozyme forms of human lactate dehydrogenase. Proteins 43, 175–185.

Saad, L.O., Mirandola, S.R., Maciel, E.N., Castilho, R.F., 2006. Lactate dehydrogenase activity is inhibited by methylmalonate in vitro. Neurochem. Res. 31, 541–548.

Scopes, R.K., 1994. Protein Purification: Principles and practice, third ed. Springer-Verlag, New York.

Segel, I.H., 1976. Biochemical Calculations: How to solve mathematical problems in general biochemistry, second ed. Wiley, New York.

Shapiro, A.L., Vinuela, E., Maizel Jr, J.V., 1967. Molecular weight estimation of polypeptide chains by electrophoresis in SDS-polyacrylamide gels. Biochem. Biophys. Res. Commun. 28, 815–820.

Shikanai, T., 2007. Cyclic electron transport around photosystem I: Genetic approaches. Annu. Rev. Plant. Biol. 58, 199–217.

Simpson, R.J., Adams, P.D., Golemis, E., 2009. Basic Methods in Protein Purification and Analysis: A laboratory manual. Cold Spring Harbor Laboratory Press, Cold Spring Harbor, NY.

Song, S.H., Dick, B., Penzkofer, A., 2007a. Photo-induced reduction of flavin mononucleotide in aqueous solutions. Chem. Phys. 332, 55–65.

Song, S.H., Dick, B., Penzkofer, A., Hegemann, P., 2007b. Photo-reduction of flavin mononucleotide to semiquinone form in LOV domain mutants of blue-light receptor phot from *Chlamydomonas reinhardtii*. J. Photochem. Photobiol. B 87, 37–48.

Stein, W.D., 1967. The movement of molecules across cell membranes. Academic Press, New York.

Studier, F., 2005. Protein production by auto-induction in high density shaking cultures. Protein. Expr. Purif. 41, 207–234.

Studier, F., Rosenberg, A., Dunn, J., Dubendorff, J., 1990. Use of T7 RNA polymerase to direct expression of cloned genes. Methods. Enzymol. 185, 60–89.

Sucharitakul, J., Phongsak, T., Entsch, B., et al., 2007. Kinetics of a two-component p-hydroxyphenylacetate hydroxylase explain how reduced flavin is transferred from the reductase to the oxygenase. Biochemistry 46, 8611–8623.

Sugiyama, N., Taniguchi, N., 1997. Evaluation of the role of lactate dehydrogenase in oxalate synthesis. Phytochemistry 44, 571–574.

Switzer, R.L., Garrity, L.F., 1999. Experimental Biochemistry, third ed. W. H. Freeman, New York.

Taguchi, H., Ohta, T., 1991. D-lactate dehydrogenase is a member of the D-isomer-specific 2-hydroxyacid dehydrogenase family: Cloning, sequencing, and expression in *Escherichia coli* of the D-lactate dehydrogenase gene of *Lactobacillus plantarum*. J. Biol. Chem. 266, 12588–12594.

Tijerina, A.J., 2004. The biochemical basis of metabolism in cancer cachexia. Dimens. Crit. Care. Nurs. 23, 237–243.

Towbin, H., Staehelin, T., Gordon, J., 1979. Electrophoretic transfer of proteins from polyacrylamide gels to nitrocellulose sheets: Procedure and some applications. Proc. Natl. Acad. Sci. USA. 76, 4350–4354.

Tsoi, S.C., Li, S.S., 1994. The nucleotide and deduced amino-acid sequences of a cDNA encoding lactate dehydrogenase from *Caenorhabditis elegans*: The evolutionary relationships of lactate dehydrogenases from mammals, birds, amphibian, fish, nematode, plants, bacteria, mycoplasma, and plasmodium. Biochem. Biophys. Res. Commun. 205, 558–564.

Tulchin, N., Ornstein, L., Davis, B., 1976. A microgel system for disc electrophoresis. Anal. Biochem. 72, 485–490.

Vallejos, R.H., Ceccarelli, E., Chan, R., 1984. Evidence for the existence of a thylakoid intrinsic protein that binds ferredoxin-NADP+ oxidoreductase. J. Biol. Chem. 259, 8048–8051.

Velick, S.F., 1949. The interaction of enzymes with small ions: An electrophoretic and equilibrium analysis of aldolase in phosphate and acetate buffers. J. Phys. Colloid. Chem. 53, 135–149.

Voet, D., Voet, J.G., 2010. Biochemistry, fourth ed. Wiley Press, Hoboken, NJ.

Voet, D., Voet, J.G., Pratt, C.W., 2013. Fundamentals of Biochemistry: Life at the molecular level, fourth ed. Wiley Press, Hoboken, NJ.

Wang, X., Ren, H., Liu, D., et al., 2013. H(+)-ATPase-defective variants of *Lactobacillus delbrueckii* subsp. *bulgaricus* contribute to inhibition of postacidification of yogurt during chilled storage. J. Food. Sci. 78, M297–M302.

Watson, J.D., 2008. Molecular Biology of the Gene, sixth ed. Pearson/Benjamin Cummings, San Francisco, CA.

Weber, K., Osborn, M., 1969. The reliability of molecular weight determinations by dodecyl sulfate-polyacrylamide gel electrophoresis. J. Biol. Chem. 244, 4406–4412.

Wetlaufer, D.B., 1962. Ultraviolet spectra of proteins and amino acids. Adv. Protein. Chem. 17, 303–390.

White, J., Hackert, M., Buehner, M., et al., 1976. A comparison of the structures of apo dogfish M4 lactate dehydrogenase and its ternary complexes. J. Mol. Biol. 102, 759–779.

Williams, S.A., Slatko, B.E., McCarrey, J.R., 2007. Laboratory investigations in molecular biology. Jones and Bartlett, Sudbury, MA.

Winfrey, M.R., Rott, M.A., Wortman, A.T., 1997. Unraveling DNA: Molecular biology for the laboratory. Prentice-Hall, Upper Saddle River, NJ.

Zanetti, G., 1981. The reduction of iodonitrotetrazolium chloride by ferredoxin-NADP+ reductase: A new tool for the characterization of the spinach chloroplast flavoprotein. Plant. Sci. Lett. 23, 55–61.

Zanetti, G., Arosio, P., 1980. Solubilization from spinach thylakoids of a higher molecular weight form of ferredoxin-NADP+ reductase. FEBS Lett. 111, 373–376.

Zeghouf, M., Fontecave, M., Macherel, D., Covès, J., 1998. The flavoprotein component of the *Escherichia coli* sulfite reductase: Expression, purification, and spectral and catalytic properties of a monomeric form containing both the flavin adenine dinucleotide and the flavin mononucleotide cofactors. Biochemistry 37, 6114–6123.

Zenno, S., Saigo, K., 1994. Identification of the genes encoding NAD(P)H-flavin oxidoreductases that are similar in sequence to *Escherichia coli* Fre in four species of luminous bacteria: *Photorhabdus luminescens, Vibrio fischeri, Vibrio harveyi*, and *Vibrio orientalis*. J. Bacteriol. 176, 3544–3551.

Zhang, H., Whitelegge, J.P., Cramer, W.A., 2001. Ferredoxin:NADP+ oxidoreductase is a subunit of the chloroplast cytochrome b6f complex. J. Biol. Chem. 276, 38159–38165.

Index

Note: Page numbers with "f" denote figures; "t" tables.

Printed and bound by CPI Group (UK) Ltd, Croydon, CR0 4YY

03/10/2024

01040318-0016